The Extinction of *Sivapithecus*

Faunal and Environmental Changes

Surrounding the Disappearance of a Miocene

Hominoid in the Siwaliks of Pakistan

AMERICAN SCHOOL OF PREHISTORIC RESEARCH MONOGRAPH SERIES

The American School of Prehistoric Research (ASPR) Monographs in Archaeology and Paleoanthropology present a series of documents covering a variety of subjects in the archaeology of the Old World (Eurasia, Africa, Australia, and Oceania). This series encompasses a broad range of subjects—from the early prehistory to the Neolithic Revolution in the Old World, and beyond including: hunter-gatherers to complex societies; the rise of agriculture; the emergence of urban societies; human physical morphology, evolution and adaptation, as well as; various technologies such as metallurgy, pottery production, tool making, and shelter construction. Additionally, the subjects of symbolism, religion, and art will be presented within the context of archaeological studies including mortuary practices and rock art. Volumes may be authored by one investigator, a team of investigators, or may be an edited collection of shorter articles by a number of different specialists working on related topics.

The Extinction of *Sivapithecus*

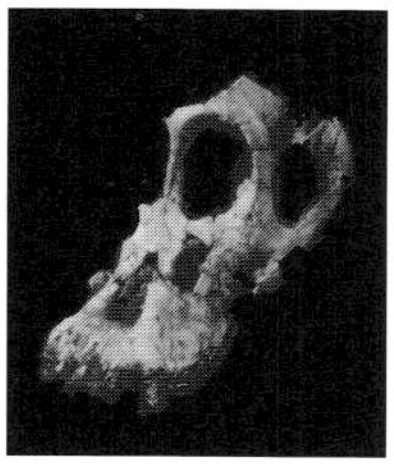

Faunal and Environmental Changes Surrounding the Disappearance of a Miocene Hominoid in the Siwaliks of Pakistan

Sherry V. Nelson

Brill Academic Publishers, Inc.
Boston • Leiden
2003

Library of Congress Cataloging-in-Publication Data

Nelson, Sherry V.
The extinction of Sivapithecus : faunal and environmental changes surrounding the disappearance of a miocene hominoid in the Siwaliks of Pakistan / Sherry V. Nelson.
p. cm. — (American schools of prehistoric research monograph series ; vol. 1)
Includes bibliographical references and index.
ISBN 0–391–04207–6
1. Sivapithecus—Siwalik Range. 2. Paleoecology—Siwalik Range. 3. Animal remains (Archaeology)—Siwalik Range. 4. Teeth, Fossil—Siwalik Range. 5. Siwalik Range—Antiquities. I. Title. II. Series.

QE882.P7N45 2003
569.9—dc21

2003051916

ISSN 1543–0529

ISBN 0–391–04207–6

PRINTED IN THE UNITED STATES OF AMERICA

The Extinction of *Sivapithecus*

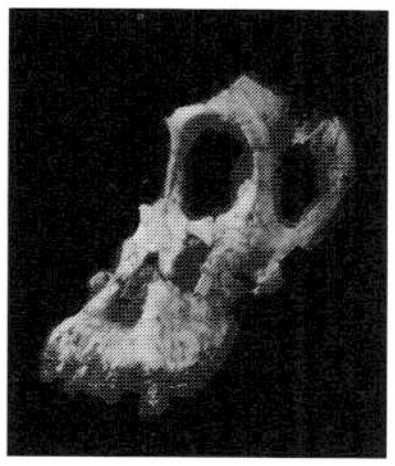

Faunal and Environmental Changes Surrounding the Disappearance of a Miocene Hominoid in the Siwaliks of Pakistan

Sherry V. Nelson

Brill Academic Publishers, Inc.
Boston • Leiden
2003

Library of Congress Cataloging-in-Publication Data

Nelson, Sherry V.
The extinction of Sivapithecus : faunal and environmental changes surrounding the disappearance of a miocene hominoid in the Siwaliks of Pakistan / Sherry V. Nelson.
p. cm. — (American schools of prehistoric research monograph series ; vol. 1)
Includes bibliographical references and index.
ISBN 0–391–04207–6
1. Sivapithecus—Siwalik Range. 2. Paleoecology—Siwalik Range. 3. Animal remains (Archaeology)—Siwalik Range. 4. Teeth, Fossil—Siwalik Range. 5. Siwalik Range—Antiquities. I. Title. II. Series.

QE882.P7N45 2003
569.9—dc21

2003051916

ISSN 1543–0529

ISBN 0–391–04207–6

PRINTED IN THE UNITED STATES OF AMERICA

Contents

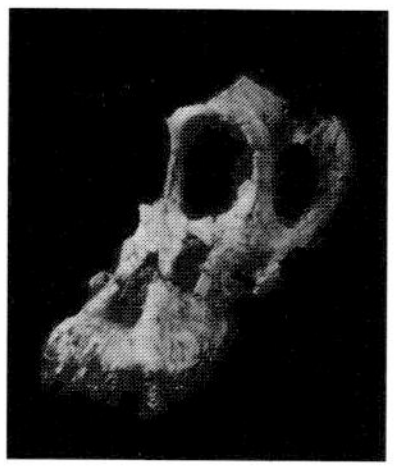

List of Figures

List of Tables

Foreword

This monograph was developed from an outstanding Ph.D. thesis, and focuses on vertebrate species from the extensive and continuously fossiliferous Miocene continental sediments of the Siwalik Series in Pakistan. Fossils from Pakistan span an interval from over 20 million years ago to less than 5 million, although this study analyzes material from the later part of that sequence. The Siwaliks are well known because of the fossil ape *Sivapithecus*, found there between 13 and 8 million years ago (although *Sivapithecus* fossils make up only a tiny fraction of the total faunal remains). Of considerable interest has been the biological adaptations of this ape, and reasons for its local extinction 8 million years ago.

Dr. Nelson's approach involves careful analyses of the faunas from two stratigraphic levels, one just before and the other just after the extinction of *Sivapithecus*. The analyses are aimed at reconstructing, through isotopic and occlusal wear studies, the diets and habitats of the fossil species, with the goal of building a detailed ecological picture of the two animal communities spanning the disappearance of the ape.

This is an unusual and very important study in that it uses a range of analytical techniques to reconstruct ecologically significant behaviors of fossil species, and does so with large samples of well dated material. Its conclusions as to the likely factors involved in the disappearance of *Sivapithecus* have general implications for the study of ape and human evolution.

David Pilbeam

Preface

The purpose of this investigation is to determine the preferred habitats of *Sivapithecus* and whether changes in habitat or climate may have led to its extinction. This research incorporates both dental microwear and carbon and oxygen stable isotopic analyses from a wide range of mammalian species in the Miocene Siwalik sediments of Pakistan to reconstruct and compare paleohabitats, climates, and dietary adaptations just before and just after the extinction of *Sivapithecus*. Habitat and climate reconstructions address forest type, mosaic of vegetation on the landscape, fruit availability, and seasonality of the precipitation regime.

The combination of both a new method of dental microwear analysis as well as stable isotopic analyses provides the most detailed reconstruction of hominoid paleoecology to date. Research presented in this volume suggests that *Sivapithecus*, unlike modern great apes, inhabited a monsoon forest component of a vegetation mosaic that also included woodland and open habitats. Nonetheless, this monsoon forest supported many species, including *Sivapithecus*, which were as dependent on fruit as their modern counterparts in rainforests. Therefore, Miocene monsoon forests probably differed from modern monsoon forests in the spatio-temporal availability of fruit, and they may have best resembled river basin forests like the Amazon today. These forests were replaced but not completely eliminated by more open habitat including patches of C_4 grass by the time *Sivapithecus* and many other frugivores and forest species went extinct, suggesting habitat fragmentation and loss of forest as likely causes of extinction.

Acknowledgments

I have many people to thank for their advice, collaboration and camaraderie throughout this research. First, I am greatly indebted to David Pilbeam, John Barry, and Mark Leighton at Harvard University for many stimulating and productive meetings that led to the initial design of this project. I am also indebted to John Barry for assistance with fossils as well with the Siwalik database.

I performed all isotopic analyses at the University of Utah in Thure Cerling's laboratory and in Jim Ehleringer's S.I.R.F.E.R. mass spectrometry facility. I especially thank Thure Cerling for introducing me to isotopic research. Benjamin Passey and Craig Cook provided critical laboratory support and I thank everyone at S.I.R.F.E.R. for their help, expertise, and generosity.

Nikos Solounias at the New York Institute of Technology graciously shared a new microwear technique that he developed, as well as his modern ruminant microwear database. Samantha Cook and Don Rubin from Harvard University's statistics department provided critical collaboration in analyzing my microwear results. Nikos, Sam, and Don Rubin were essential to designing the new microwear method presented in this book.

Numerous people provided advice and training within their fields of expertise. Ray Bernor from Howard University taught me equid tooth identification and provided the Ethiopian and German teeth sampled in this project. Felicitas Weidemann and Dan Lieberman, then at George Washington University, shared with me their knowledge and laboratory skills in studying tooth histology.

I would also like to thank everyone who assisted me with fieldwork in Pakistan, including Kay Behrensmeyer, Catherine Badgley, Nina Jablonski, Jason Head, Michele Morgan, members of the Geological Survey of Pakistan, and most especially John Barry.

This research was generously funded by the National Science Foundation, the American School of Prehistoric Research, the Mellon Foundation, and the Packard Foundation.

Finally, I thank everyone in the Department of Anthropology at Harvard University for providing a stimulating academic community throughout this research, including Catherine Smith, Michele Morgan, Lawrence Flynn, Richard Wrangham, Nathan Young, Michael Steiper, and Martin Muller. I also thank Wren Fournier for editing this volume with unfailing energy and enthusiasm.

1 Introduction

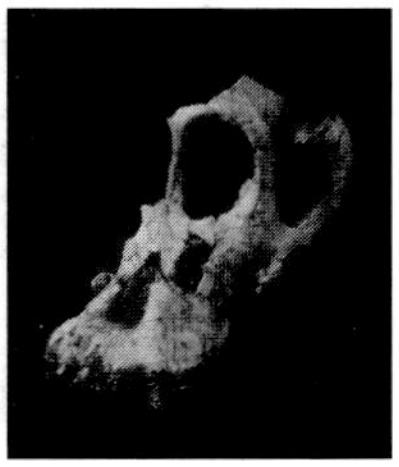

Synopsis

The Miocene was a period of dramatic evolutionary change within the hominoid clade, including the appearance of the first undisputed ape species during the early Miocene, followed by a radiation of the clade throughout the epoch, resulting in a diversity of species inhabiting Africa, Europe, and Asia. By the end of the Miocene, however, most members of the clade had become extinct, with the geographic range of hominoids greatly reduced. Isotopic, botanical, and faunal evidence indicate that, due to development of the polar cryosphere, the second half of the Miocene was marked by both global cooling and aridification in middle and low latitudes, with closed forest being replaced in many areas by more open habitat (Barry 1995; Kennett 1995; Quade et al. 1995a; Quade and Cerling 1995). Only the Siwalik sediments of Pakistan, spanning the past 20 million years, provide a nearly continuous record of faunal and environmental change documenting an ape clade throughout its radiation and demise (see Figures 1.1 and 1.2). Within this long terrestrial sequence, the hominoid genus *Sivapithecus* appears around 13 Ma and disappears around 8.4 Ma (Barry et al. 2002).

The purpose of this investigation is to determine what roles changing habitat and climate, including increasing seasonality, may have played in the extinction of *Sivapithecus*. This research incorporates both dental microwear and carbon and oxygen stable isotopic analyses from a wide range of mammalian species to reconstruct and compare paleohabitats, climates, and dietary adaptations just before and just after the extinction of *Sivapithecus*. Habitat and climate reconstructions address forest type, mosaic of vegetation types on the landscape, fruit availability, and degree of seasonality of the precipitation regime. Microwear and isotopic analyses are then used to infer what changes in habitat or rainfall patterns preceded the extinction of *Sivapithecus*. Finally, the inferred habitat requirements of *Sivapithecus* will be compared to those of modern apes with the aim of better understanding how *Sivapithecus*, and possibly most other Miocene hominoids, may have differed ecologically from extant species.

Background

Modern Ape Habitats

In order to interpret the effects of changing environment on fossil hominoids, it is first necessary to understand the habitat and climatic requirements of modern apes. Extant apes are found only in Africa and southeast Asia, and with few exceptions, their geographic ranges lie within a few degrees of the equator. Within the equatorial region, precipitation regimes are such that dry periods generally last no longer than four months. Length of the dry period can have important consequences for hominoids because all, except the mountain gorilla, have diets specializing in ripe fruits, and fruit temporal availability, density, and diversity often correlate negatively with length of dry seasons.

Within tropical Africa, availability of fruit varies seasonally, with fruit peaks coinciding with annual rainfall peaks in most regions, though interannual variation in rainfall does not correlate with "good" fruiting years versus "bad" (Tutin et al. 1991; Wrangham et al. 1993). Equatorial west and central Africa experience two short relatively wet seasons and two short relatively dry ones annually, but even during the drier months, most regions remain humid (Walter et al. 1975).

The vegetation of west and central equatorial Africa consists of evergreen forest and is inhabited by lowland gorillas, chimpanzees, and bonobos. The Lope Reserve in Gabon, where gorillas and chimps live sympatrically, experiences four three-month seasons (Hladik 1977), and annual variability in fruit abundance correlates with seasonality of rainfall, with maximum fruit availability during the major rainy period (Tutin et al. 1991). Furthermore, the mean number of fruit species eaten by chimps and gorillas correlates positively with variability in fruit phenology, though even in drier months and in "bad" fruiting years, both species are still able to find fruit (Tutin et al. 1991).

Bonobos in Lomako Forest in Zaire experience a similar seasonal regime as apes in Lope, with four seasons, and again, fruit diversity fluctuates with respect to season. Diversity drops during the driest season and low diversity then continues into the minor wet season. Again, bonobos are never without ripe fruit, however. During a 19-month study by Badrian and Malenky (1984), number of available fruit species known to be eaten by bonobos never fell below seven for any given month.

Among the African apes, only chimpanzees range outside tropical forests. In east Africa, due to the Western Rift mountain ridges acting as a barrier to rain-bearing western winds, the bimodal rainfall pattern is characterized by unequal peaks, namely a long rainy season and a short rainy season (Kortlandt 1983). Climate is no longer

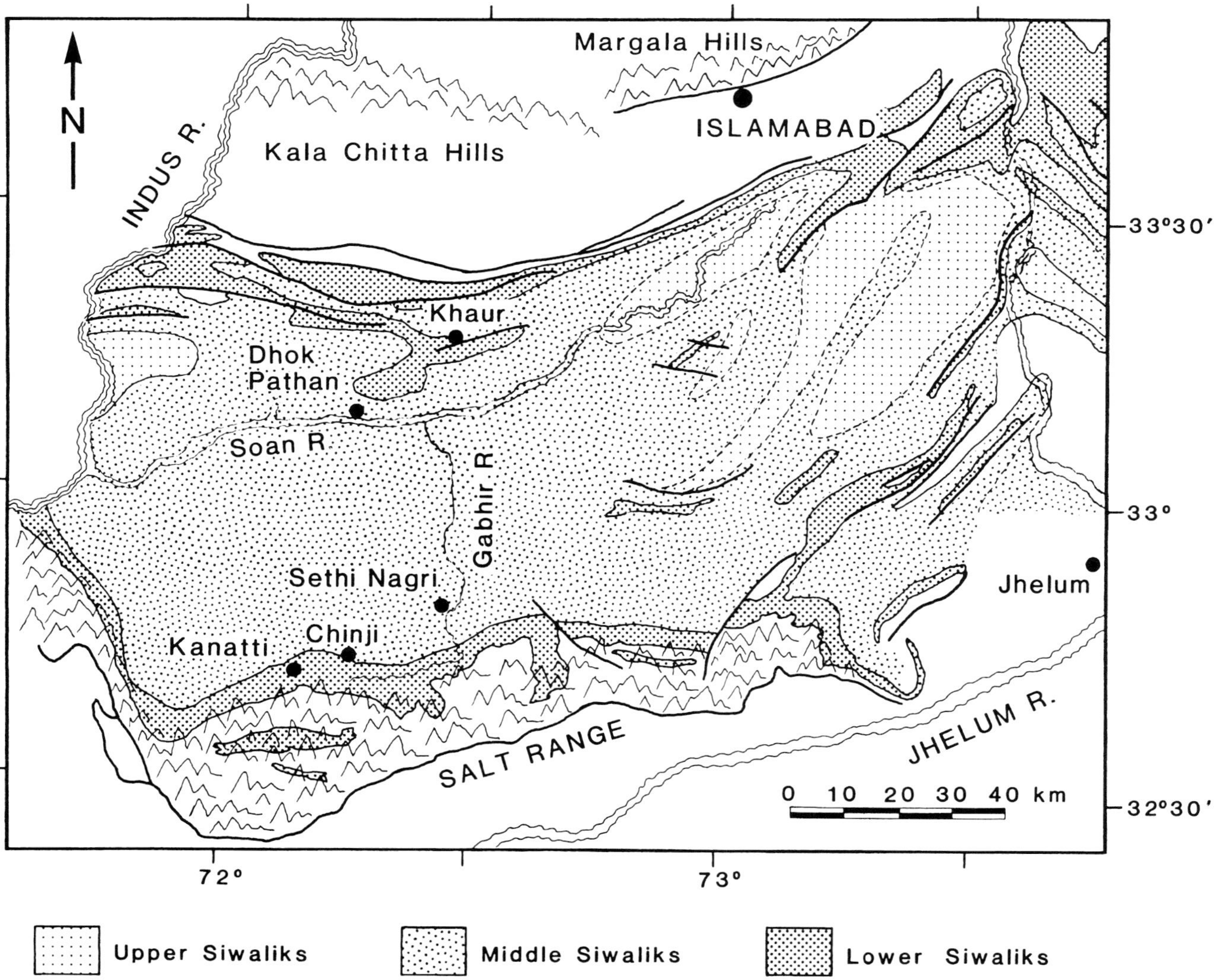

Figure 1.1 The Siwalik sediments of northern Pakistan span hundreds of kilometers across the Potwar Plateau. Deposits and formations are named after nearby towns. Map copyright and courtesy of C. E. Badgley and *Palaeogeography, Palaeoclimatology, Palaeoecology*.

continuously humid, and a longer dry season results in less evergreen and more semi-evergreen and deciduous forests in east Africa (Walter and Breckle 1984; Walter et al. 1975). The chimpanzee site of Kibale in Uganda is a mosaic of moist forest (comprising 60 to 70 percent of the area), tall grassland, swamps, and softwood plantations (Wrangham et al. 1993, 1991). While Kibale experiences two annual rainy seasons, it shows no correlation between fruit abundance and the annual cycle of rainfall (Wrangham et al. 1993), though according to Malenky and Wrangham (1994), fruit scarcity occurs when rainfall is high. Consumption of pith does, however, positively correlate with monthly rainfall and negatively with fruit abundance, suggesting chimps eat pith as a fallback source of carbohydrates when fruit is scarce and when pith is most nutritious, during periods of growth when rainfall is high (Malenky and Wrangham 1994). Again, though, chimpanzees in east African sites are never without ripe fruit. A two-year study of chimpanzee diets at Gombe found that percentage of fruit in the diet never fell below about 45 percent (Goodall 1986).

Chimpanzees also live in regions that experience extensive dry periods of five months or more, although in these regions, such as Mt. Asserik in Senegal, population density is extremely low, and home ranges are large. According to a four-year study by McGrew et al. (1981), for five consecutive months of the year, rainfall averaged less than 100 mm, most of which fell in one month. Mean annual rainfall fell between 824–1,220 mm, compared to averages of around 1,500 mm for most other chimp sites. Baldwin et al. (1982) estimated population density to be 0.09 chimp/km^2, comparable to Ugalla, another marginal habitat site (Kano 1972). These densities are much lower than those of chimps at Gombe or Kasoge, where density is 1.1–2.5 chimps/km^2 (Baldwin et al. 1982; Goodall 1968; Wrangham 1975; Nishida and Kawanaka 1972). The density difference lies in home range sizes, with estimated range for Mt. Asserik chimps being 278–333 km^2, compared to no more than 40 km^2 for chimps at Gombe and Kasoge (Baldwin et al. 1982). While almost all rainfall is limited to five months of the year, predictability of rainfall is very high (McGrew et al. 1981), and while chimps travel freely during the wet season to reach dispersed fruit, during the dry season, their ranges are confined primarily to gallery forest (Baldwin et al. 1982). Thus, while chimpanzees inhabit regions with extensive dry seasons, their population size is extremely low, and during the dry season, they are forced into habitat similar to typical hominoid sites, namely wet forest.

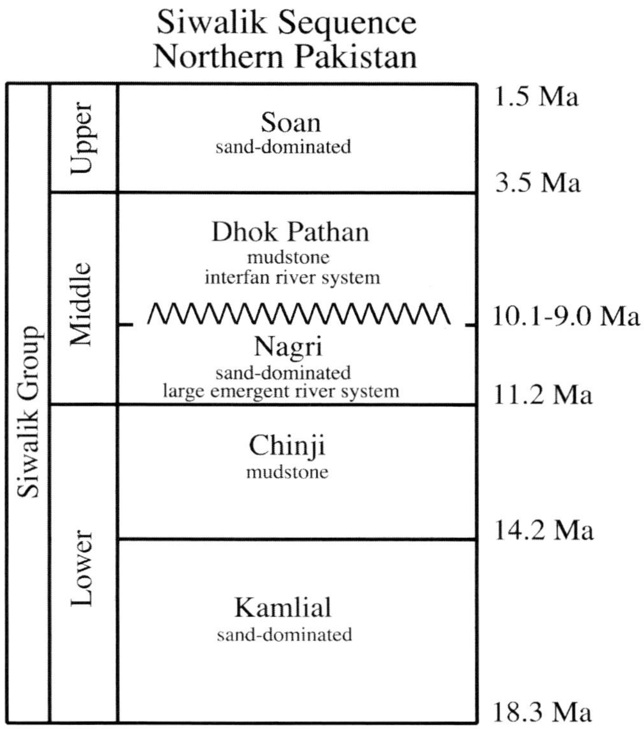

Figure 1.2 Formations of the Siwaliks are characterized by kilometer-scale alternations in lithologies that represent changes in channel belts over time. Overlap in ages between the Nagri and Dhok Pathan represents time transgression.

Southeast Asian dipterocarp rainforests differ dramatically from African ape habitats in phenology, with rainforests characterized by mast fruiting events in which a large number of trees fruit in synchrony about every two-to-ten years (Knott 1998). Time of fruiting is therefore very irregular rather than following an annual seasonal regime. Rainfall regimes are highly seasonal for these forests, however, due to the monsoonal system. Monsoonal rainfall patterns influence vegetation patterns in that forest type, such as evergreen versus deciduous forest, is largely determined by length of the dry season as well as total annual rainfall (Walter 1973). The boundary between rainforest and monsoon forest formations is determined by evapotranspiration and availability of soil water and is therefore most closely related to length and severity of the dry season rather than total annual rainfall (Richards 1996). Most evergreen rainforests are limited to regions that experience only a very short dry season of two months, if any dry season occurs at all. As length of dry season increases and annual rainfall amount decreases, a transition to moist deciduous and then to dry deciduous forest occurs. For example, western Java, covered by rainforest, experiences 3,000–4,000 mm rain/year, and over 100 mm rain falls every month. Eastern Java, on the other hand, receives only 1,300–2,500 mm rain/year and experiences a drought of two-to-four months. As a consequence, the vegetation of this portion of the island is moist deciduous forest. As length of dry season increases such as in parts of south China, which experience four-to-six month droughts, the vegetation type becomes dry deciduous forest (Walter and Breckle 1984).

Among this spectrum of forest types, siamangs and orangutans are limited to rainforests, with only gibbons ranging into more seasonal forests, although gibbons' geographic range lies predominantly in rainforest as well. Both geographic range and day range among the southeast Asian hominoids correlate negatively with body size, with gibbons having half the body size but twice the day range of siamangs, and orangutans having the smallest day range. Gibbons travel greater daily distances in order to exploit small, scattered fruit patches, while siamangs travel half as far and feed at half as many sites, depending instead upon clumps of highly productive trees (Chivers 1984; MacKinnon and MacKinnon 1978). Orangutans have the greatest dependence on high average ripe fruit abundance, but they are able to store large amounts of energy as fat and therefore take advantage of periods of fruit superabundance, something not yet observed in gibbons (Knott 1998; MacKinnon 1977). Perhaps as a result of differences in fruit density needs, gibbons have the widest geographic distribution among the southeast Asian hominoids and are found as far west as Assam and Bangladesh (Gittens and Tilson 1984). In both Malaysia and Sumatra, where gibbons and siamangs live sympatrically, gibbons have the wider distribution. In Sumatra, where all three groups occur, the distribution of orangutans, the species most dependent on highly productive clumps of fruit, lies entirely within that of siamangs (MacKinnon 1977). In Segama, Borneo, where gibbons and orangutans live sympatrically, an orangutan might travel 0.5 km/day, while a gibbon can travel several kilometers each day. As a result, even when fruit is scarce and orangutans are eating bark and leaves, gibbons are able to maintain a high proportion of fruit in their diets (MacKinnon 1977).

While gibbons do have the widest range, even they appear to be restricted to the more diverse and less seasonal forests. In Java, gibbons are only found in rainforests on the western half of the island. No gibbons are found in the deciduous forests on the eastern half (Kappeler 1984). In Kha Khaen, Thailand, lar gibbons are able to thrive in a mosaic of semi-evergreen and mixed deciduous forests where there is a dry season of several months, but they avoid the drier forests where there is low plant species diversity. Chivers (1984) suggests that gibbons prefer the environments with high tree diversity because these habitats buffer the effects of seasonality by enabling animal-dispersed fruit to be available throughout the year. Thus while fruit abundance and density decreases during the dry months, a high diversity of fruiting trees can ameliorate the effects of low abundance in any one preferred fruiting species.

Frugivorous primates thus show a number of different adaptations for dealing with the spatio-temporal variability of fruit. Cercopithecines, which feed on fruit seeds, lengthen their effective fruiting season by eating unripe fruit, having developed a tolerance to their toxins (Wrangham et al. 1998). This adaptation also allows them access to fruit sooner than apes, which have a far lower tolerance to toxins. Some apes increase daily travel distance in order to find more fruit patches. This adaptation requires efficient locomotor adaptations such as brachiation in the lesser apes (Cannon and Leighton 1996, 1994) and terrestrial adaptations such as knuckle-walking in African great apes. The orangutan, with neither of these locomotor adaptations, is able to take advantage of masts by fat storage. All apes rely on fallback foods during fruit shortages, with larger body size being an advantage, but the question remains-how long a period can a frugivorous ape rely on fallback foods? The evidence from Africa would suggest about three or four months for almost all populations. Finally, another option might be reduction in energy expenditure during lean times, but given the great ape life history pattern, a significant reduction may not be possible, at least for lactating females. Knott (1998) has found that nursing females incur a significant energy deficit during periods of low fruit availability.

Thus, modern frugivorous ape habitats are restricted to regions characterized by the following: 1) the presence of at least some gallery/rainforest; 2) greater than 1,000 mm of annual rainfall, and usually at least 1,500 mm of rainfall; 3) fruit available year round, with lean periods lasting no more than four months, unless the ape has fat storage capabilities, and; 4) availability of fallback foods, including species which fruit year-round, such as *Ficus*.

Ficus fruits are important fallback foods primarily because of their aseasonal phenology and enormous crop sizes (Lambert 1989; Leighton 1993). They are not preferred fruits, however, due to nutritionally inferior pulp, being very low in digestible carbohydrate, protein, and fat, and relatively high in tannins and fiber content (Leighton 1993).

Sivapithecus

While *Sivapithecus* was similar to modern hominoids in some aspects of its biology, it also shows some significant differences that may have resulted in very different habitat requirements, including deviations from the fruit availability requirements listed above. *S. sivalensis* males were about the size of male chimpanzees (40–50 kg), while females were roughly the size of male baboons (20 kg). *S. parvada* is the larger species, with males approximating male orangutans or female gorillas (Kelley 1986). The molars of *Sivapithecus* have low, rounded cusps and hyperthick enamel which wears to a relatively flat occlusal surface with no dentine pits, a morphology most similar to extant primates that feed on hard objects, such as fruits with tough rinds or nuts (Kay 1981). However, preliminary microwear analyses of *Sivapithecus* indicated a diet most closely resembling that of chimpanzees rather than that of hard-object feeders (Teaford and Walker 1984). *Sivapithecus* craniofacial morphology is heavily buttressed, suggesting an adaptation to either withstand prolonged cyclical loading or to generate high occlusal loads (Kelley and Pilbeam 1986). The transversely thick mandibular corpus would withstand large twisting moments, and the deep symphysis might counteract dorsoventral bending. The mandibular ramus is both anteroposteriorly deep and high, indicating large moment arms for large temporalis and masseter muscles and thus a high bite force. An anteriorly placed zygomatic also suggests a masseter capable of generating high masticatory forces. Given that *Sivapithecus* lacked molar shearing crests of folivorous primates and that the molars most resemble those of frugivores, it is likely that these adaptations for heavy mastication may have related to fallback foods and that *Sivapithecus* diet consisted primarily of ripe fruit (Kelley and Pilbeam 1986).

The postcranial skeleton of *Sivapithecus* suggests that it was a pronograde quadruped (similar to monkeys) like most other Miocene hominoids, but that it included some climbing in its locomotor repertoire as well (Rose 1986). The distal humerus, characterized by highly differentiated trochlea and capitulum, as well as a deep olecranon fossa, resembles that of modern apes and indicates full extension capabilities as well as a wide range of pronation-supination possible throughout extension (Kelley and Pilbeam 1986). Furthermore, the surface of the capitate which articulates with the lunate is quite curved, suggesting wide rotation of the lunate during pronation-supination. However, the capitate suggests no features related to stability of the midcarpal region during the extended and pronated position of knuckle-walking. A robust partial pollical phalanx shows well-developed attachment areas for collateral ligaments and suggests powerful flexion. The proximal femur shows anteroposterior flattening, suggesting strengthening against laterally directed forces during adduction-abduction of the hip, and extension of the articular surface of the femoral head onto the neck suggests a mobile hip joint as well. The knee is ape-like in its mobility. Features of the foot and ankle, like the hand, suggest powerful grasping as well as a stable ankle joint. A deep cuboid facet indicates a range of pronation-supination and grasping comparable to living hominoids, as does a strong groove for the flexor digitorum brevis. Again, however, the foot lacks any features related to specialized locomotion, such as the extreme mobility and quadrumanous suspension found in Pongo. Thus *Sivapithecus* was probably predominantly arboreal and above-branch, with some climbing behavior and little ape-like suspensory activity. While no features preclude terrestrial locomotion, *Sivapithecus* shows no specializations for extensive terrestriality (Kelley and Pilbeam 1986).

Finally, mandibular and premaxillary fragments of a *S. parvada* juvenile, the largest *Sivapithecus* species, indicate that this species had a life history pattern similar to that of modern great apes (Kelley 1997). This individual died four to six months after eruption of the first molar (M1), and counts of incisor perikymata indicate that the individual was about 40–41 months old when this M1 erupted, comparable to the age of M1 emergence in chimps and gorillas (approximately 39 and 42 months respectively). Furthermore, this age is substantially greater than the age of M1 eruption in any cercopithecoid (generally no more than 20 months). Given that individual life history traits are strongly correlated with the time of eruption of the M1 (Smith 1989, 1991), the juvenile specimen is a strong indication that *Sivapithecus* followed a modern great ape life history trajectory.

Thus, the morphology of *Sivapithecus* suggests that it was a large-bodied frugivore that would have been susceptible to periods of ripe fruit shortage. How, then, could *Sivapithecus* have dealt with the spatio-temporal variability of food resources? Given its body size, inability to brachiate, and lack of terrestrial specializations, it is unlikely that it could have traveled as extensively as gibbons or chimpanzees in search of fruit. With the life history pattern of a great ape, a decrease in energy expenditure alone was probably not an option, at least for lactating females. Fat reserves are an adaptation to environments which experience extremes, such as temperate regions or the masting forests of southeast Asia. Such seasonal extremes may have existed in *Sivapithecus*'s habitat, but there is no current evidence for masting. Perhaps *Sivapithecus* had a greater tolerance of toxins than modern apes, or, given the novel dietary adaptations suggested by the cranial and dental morphology, *Sivapithecus* may have relied upon different fallback foods, such as hard seeds and nuts, or may have relied upon them to a greater extent. Greater detail of *Sivapithecus*'s habitat in terms of forest type and seasonality is needed to better understand its requirements, adaptations, and extinction.

Climate and Habitat Changes in the Siwalik Sequence

The extinction of *Sivapithecus* can be placed in a context of faunal and environmental change, for the Siwalik sediments span the past 20 million years, thus documenting the clade throughout its radiation and exinction. The Siwaliks are fluvial deposits exposed throughout significant areas of western and northern Pakistan, northern India, and Nepal (see Figures 1.1 and 1.2). In Pakistan, sediments are particularly well exposed on the Potwar Plateau, where they are typically 2,000–5,000 m thick and comprise broad outcrop bands extending laterally for tens of kilometers (Barry 1995). These outcrops span 18 Ma to less than 1 Ma, making the Potwar formations one of the longest, most fossiliferous terrestrial vertebrate sequences known. While the abundance of fossils varies throughout the sequence, many subintervals are very well represented, allowing for a temporal resolution of 100,000 years in documenting patterns of faunal change. Barry et al. (1990, 1995) have used both interval-quality scores as well as an index of completeness to control for biases which could affect inferences concerning faunal patterns, such as differences in depositional environment or in prospecting and fossil collection, and they have currently added Koch's (1987) method of estimating range extensions from occurrence data as well. Having corrected for biases, Barry (1995) found statistically significant maxima of first appearances at 13.5 and 8.5 Ma, and maxima of last appearances at 12.5 and 8.0 Ma. A younger faunal turnover event between 7.3 and 7.0 Ma comprises both first and last appearances (Barry et al. 2002). Late Miocene faunal change accompanies a pattern of global cooling and aridification. Among rodents, closed-habitat species such as sciurids, glirids, and cricetids, are replaced by murids around 7.5 Ma, and high-crowned tachyoryctines replace low-crowned rhizomyines (Flynn et al. 1995; Flynn 1982; Flynn and Jacobs 1982). Among large mammals, bovids become more abundant compared to tragulids between 9.5–7.5 Ma, and body size increases in many taxa, particularly at the maximum of first appearances at 13.5 and 8.5 Ma (Barry 1995). Furthermore, many large mammalian taxa disappear, including two suids, all but one tragulid, the giraffe *Bramatherium*, a bovid lineage, and *Sivapithecus*. These taxa are replaced by true giraffes, larger and more hypsodont bovids, colobine monkeys, and more open-habitat taxa such as rabbits (Barry 1995).

Environmental changes in the Pakistan Siwalik sequence have best been documented by isotopic changes in paleosols, given the absence of a pollen record in Pakistan. Most terrestrial plants assimilate CO_2 by one of two photosynthetic pathways, and these different pathways fractionate carbon isotopes differently. C_3 plants include most trees and shrubs and some grasses which experience cool growing seasons. C_4 plants, on the other hand, are grasses and sedges that live in hot, drier environments, and these plants have tissues enriched in ^{13}C relative to C_3 plants. Paleosol $\delta^{13}C$ values reflect the isotopic composition of vegetation growing above the site (Cerling et al. 1989). When dated according to the Cande and Kent (1995) timescale, the paleosol data indicate that C_4 plants appeared around 8 Ma, with the first C_4 dominated habitats at 7.37 Ma (Quade and Cerling 1995; Quade et al. 1989). Communities comprised of predominantly C_3 plants became diminished or disappeared around 7.0 Ma.

Carbon isotope ratios in mammalian tooth enamel corroborate the paleosol data. Previous analyses of mammalian tooth enamel indicate a transition from C_3 dominated diets to mixed C_3/C_4 diets and ultimately to C_4 dominated diets, with a major shift after 8 Ma (Morgan 1994). The paleoenvironment was most likely a mosaic of C_3 and C_4 plants throughout the Late Miocene, but with increasing percentage of C_4 grasses after 8 Ma.

Changes in the vegetation mosaic probably related to changes in the depositional systems of the Pakistan Siwaliks during the Late Miocene as well as to changes in climate. Nagri Formation large emergent river systems were replaced by Dhok Pathan Formation interfan river systems beginning around 10.1 Ma (Barry et al. 2002; Willis and Behrensmeyer 1995). After 9.0 Ma, these interfan river systems were less well drained, with more seasonally variable flow and more avulsions. Vegetation probably responded to specific local conditions, resulting in mosaics over distances of hundreds of meters to a few kilometers (Barry et al. 2002).

The paleosol data also indicate a shift in $\delta^{18}O$ values, beginning around 9.15 Ma, with samples becoming more enriched over time (Quade and Cerling 1995; Quade et al. 1989). This shift has been interpreted as both a decrease in rainfall and increase in seasonality of rainfall. Foraminiferal records indicate the beginning of strong summer Arabian Sea upwellings around 9.4 Ma, with continued strengthening until around 8.2 Ma, suggesting the beginning or at least major intensification of the south Asian monsoon (Kroon et al. 1991, corrected to Cande and Kent 1995). Intensification of the monsoon may explain the oxygen shift in Pakistan, with intensification causing increased seasonality, aridification, and summer rainfall which would then favor C_4 plants and hence more open habitats, over C_3 plants and closed forest.

While no pollen has been found associated with the Pakistan Siwaliks, fossil wood in India indicates that tropical evergreen forests with rare moist deciduous taxa flourished in this region during the Middle Miocene (Prasad 1993). Though poorly dated, these forests were probably present between 17–11 Ma. Moist deciduous taxa later dominate, and dry deciduous forests make their first appearance, but again this transition is poorly dated. Pollen and leaf fossils from Nepal also suggest that evergreen forests dominated the vegetation in this region during the Middle Miocene but were gradually replaced by semi-deciduous and dry deciduous forests between 11–6 Ma (Quade et al. 1995a). Recent work by Corvinus and Rimal (2001) provides a more detailed plant record of the Nepal Siwaliks. The Bankas formation, corresponding to Pakistan's Chinji Formation (14.2–11.2 Ma) (Barry et al. 2002), preserves leaves of tropical evergreen genera found today only in southeast Asia, including *Dipterocarpus*. Moist deciduous elements make their first appearance in the lower Chor Khola, corresponding to the beginning of the Pakistan Nagri Formation at 11 Ma, though evergreen plants are still present. Evergreen taxa become rare and are replaced by a majority of moist deciduous and dry deciduous taxa in the Surai Khola, corresponding to the second half of Pakistan's Dhok Pathan Formation, roughly 7 Ma. A shift from moist to dry deciduous flora subsequently occurs in the Dobatta Formation, equivalent to the Pakistan Upper Siwaliks some time after 4 Ma. A high proportion of Nepal Middle Miocene paleosols are gleyed, suggesting swampy conditions. It should be noted that gleyed paleosols are absent from the Pakistan Siwaliks, however, indicating that Nepal was wetter than Pakistan during the Miocene, as it is today. If so, then the transition from evergreen to moist deciduous to dry deciduous forests probably occurred slightly earlier in Pakistan compared to Nepal. The Nepal paleosols also show oxygen and carbon isotope shifts comparable to those seen in Pakistan, though the carbon shift is about 0.7 million years later in Nepal (Quade et al. 1995a).

Specific Aims

The purpose of this investigation is to first determine the preferred habitat of *Sivapithecus* in terms of forest type, landscape diversity, fruit availability, and seasonality. A second goal is to then determine what changes in habitat and climate, if any, are correlated with the extinction of *Sivapithecus*. Finally, what is learned of the paleohabitats, in conjunction with morphological evidence, will be used to infer how, if at all, *Sivapithecus* may have differed from modern hominoids in diet or habitat requirements. With respect to habitat and climate, the following questions will be addressed.

I. When *Sivapithecus* was abundant, and again when *Sivapithecus* became extinct, what was the habitat like in terms of seasonal, phenological, and vegetation mosaic patterns? Was it an evergreen forest, a deciduous forest, a savanna/woodland, or is there no modern analogue, i.e. inferred seasonal, phenological, and vegetation mosaic patterns do not all suggest the same forest type? Was it similar to modern ape habitats today, i.e. could it support a large-bodied ape whose diet was dominated by ripe fruit?

II. Did rainfall patterns become more seasonal, or did the climate become more arid prior to the extinction of *Sivapithecus*? If so, then at what time, if any, did the dry season become more extensive than those modern apes experience?

III. Was the habitat of *Sivapithecus* the same as that of modern apes, and by inference, were its habitat requirements the same? If so, then its habitat should resemble those of modern apes in forest type and seasonal regime unless evidence suggests that Siwalik forests were unlike any today.

IV. If *Sivapithecus*'s habitat did resemble those of modern apes, was it ever more marginal than that experienced by apes today? Does significant habitat and/or climate change coincide with the extinction of *Sivapithecus*, suggesting a causal relationship?

Alternatively, the habitat of *Sivapithecus* may never have been like those of extant hominoids. If Siwalik forest type has a modern analogue, and this analogue does not support apes today, it suggests that *Sivapithecus*, and possibly other Miocene apes, differed significantly in some aspect from modern apes, such as less dependence on ripe fruit or dependence on very different fallback foods. On the other hand, if the habitat resembled that of modern apes even when *Sivapithecus* became extinct, it is possible that *Sivapithecus* had an even greater dependence on ripe fruit or large fruit patches. Or, the extinction of *Sivapithecus* may not have been related to habitat requirements at all. Perhaps *Sivapithecus* experienced greater predation pressures than modern apes, such as pressure from sabertooth cats. Increasingly open habitat may have increased these pressures. Given that *Sivapithecus* was one of many hominoid taxa through Europe and Asia to go extinct in the Late Miocene, disease is not likely to have been the culprit. Finally, if competition were a factor, then unless it was for fallback foods, there should have been an increase in numbers of frugivores or new frugivore taxa, which could be detected in this study (though no previous study has given any indication of this possibility).

Finally, it is also possible that Siwalik forests had no modern analogue. This scenario could be detected if inferred seasonality of precipitation, fruit availability, and vegetation mosaic do not all indicate the same forest type. It is possible that Miocene forests differed in their spatio-temporal availability of fruits compared to modern forests due to differences in plant species, climate, or atmospheric conditions that might affect plant productivity, abundance, or diversity.

Outline of Analyses

Chapter Two

Chapter Two presents dental microwear analyses of a wide range of mammalian taxa, including *Sivapithecus*. The purpose of this investigation is to determine the preferred diet of *Sivapithecus*, fruit availability in its habitat, and any changes in fruit availability that might have coincided with its extinction. A new light microscope technique is used which allows for analysis of many specimens, and a

Bayesian approach to statistical analyses of the data allows dietary results to reflect degree of similarity to multiple modern species or diets, permitting the possibility that there is no perfect modern analogue to a fossil diet. Furthermore, fossil diets can be interpreted in terms of how much each food type contributed to the diet and how those contributions may have changed over time.

Results indicate that *Sivapithecus* was frugivorous, as were many other contemporaneous taxa, including suids, tragulids, and some bovids. Fruit availability appears to decrease around the time *Sivapithecus* became extinct, with fruit replaced by browse in some taxa, and browse replaced by C_4 graze in others, indicative of forest replacement by more open habitat. However, persistence of frugivorous suids, tragulids, and very small bovids after the disappearance of *Sivapithecus* suggests that moist forests did not disappear entirely, but likely became too fragmented to support a large-bodied frugivorous ape.

Chapter Three

Chapter Three presents an isotopic study of a wide range of mammalian taxa for two time intervals—9.3–9.2 Ma, when *Sivapithecus* was present, and 8.1–8.0 Ma, shortly after *Sivapithecus* became extinct. The purpose of this investigation is to reconstruct the vegetation mosaic at both time levels in order to determine what *Sivapithecus* habitat was like and what changes in habitat and climate may have taken place by the time *Sivapithecus* went extinct. Both carbon and oxygen stable isotopes were sampled from inorganic carbonate in tooth enamel. Results indicate a vegetation mosaic of both closed and open habitat at both time intervals, but a decrease in forest accompanied by an increase in open habitat (including C_4 grasses) during the younger interval. Individuals from many species cluster with each other with respect to both carbon and oxygen isotope values, indicating a spectrum of feeding adaptations exploiting different parts of the habitat. Isotopic values for *Sivapithecus* suggest that it fed in the forest upper canopy. Taxa feeding in the most closed, wet habitat became extinct by 8.1 Ma. Furthermore, enrichment of oxygen isotope values suggests a change in climate with a reduction in annual rainfall, perhaps accompanied by changes in precipitation sources or rainfall regime as well. These results suggest that forests became fragmented over time. While still present in the younger level, much of the forest was replaced by open habitat, including patches of C_4 grass. Forest fragmentation is a likely cause of the extinction of *Sivapithecus*.

Chapter Four

Chapter Four presents a study of intra-tooth isotopic variability measured in hypsodont equid teeth as a means of determining paleoseasonality. The purpose of this investigation is to determine whether *Sivapithecus* habitats were characterized by a seasonal rainfall regime, and if so, what kind of forest is associated with this regime. Forest type, such as evergreen, semi-evergreen, moist deciduous, and dry deciduous, is largely determined by length of the dry period. Reconstruction of the rainfall regime therefore allows inferences of forest type in the fossil record. Seasonal rainfall regimes can be determined from oxygen stable isotopes, for oxygen isotopes track amount of rainfall in tropical regions. Sampling oxygen isotopes along the length of an equid tooth, which takes over a year to develop, therefore provides a record of seasonal isotopic changes in body water, which in turn reflect drinking water and ultimately precipitation. Results from equid teeth spanning 10.0–6.3 Ma indicate a rainfall regime similar to that experienced by southern China monsoon forests today, with a decrease in rainfall over time. These results suggest that *Sivapithecus* was living under a seasonal regime with dry periods of five to six months, longer than the maximum of four months that most great apes experience today. Either *Sivapithecus* differed in habitat requirements, or the forest differed in terms of the spatio-temporal variability of fruit compared to modern analogues.

Equid teeth analyzed in this study include premolars and molars. When oxygen values are used to determine the season in which each tooth terminated crown formation, a seasonal birthing pattern is strongly suggested. It appears that most Siwalik horses were born during the wet season, like Burchell's zebra today (Nowak 1999).

This study tracks carbon as well as oxygen isotopic variability in the same teeth. The first indisputable evidence for C_4 graze in the diet occurs at 8.7 Ma in the Siwaliks, with the first evidence of C_4 graze anywhere so far discovered in 10.8 Ma Ethiopian equid teeth. Graze becomes more prevalent around 8.0 Ma, coinciding with the first C_4 evidence from Siwalik paleosols. Even as late as 6.3 Ma, however, browse continued to be a major component of equid diets year-round.

Chapter Five

Chapter Five provides a synthesis of microwear and isotopic results per fossil taxa studied in these investigations in order to better infer diets and paleohabitats. The benefit of combining microwear and isotopic analyses is that each method compensates where the other fails. Without corroborative isotopic analyses, microwear resulting from hard-shelled fruits and seeds cannot be distinguished from wear resulting from C_4 phytoliths. On the other hand, while isotopic analyses provide quantifiable differences between feeding on C_3 versus C_4 plants and between feeding in closed versus open habitat, they do not allow finer distinctions between C_3 browsers and C_3 frugivores. This synthesis indicates that Late Miocene fauna included many frugivores as well as open-habitat browsers and eventually mixed C_3/C_4 feeders. Preferred species habitats and diets correlate with times of extinction, with closed-habitat taxa and frugivores going extinct first and open-habitat taxa prevailing until later in the record.

Isotopic results from the mammalian tooth enamel analyses are then compared to previous isotopic analyses of paleosol carbonates. Analyses of mammalian teeth detect the first evidence of C_4 grasses, but both mammalian teeth and paleosol analyses detect significant patches of C_4 within the vegetation mosaic around 8.0 Ma. Both methods also detect a shift in oxygen values beginning around 9.3–9.2 Ma, followed by another shift around 8.0 Ma. With respect to the vegetation mosaic, analyses of mammalian teeth detected a wider range of habitats at each level studied compared to paleosols, but the paleosol sample size was small. The benefit of combining mammalian teeth and paleosol isotopic analyses is that they provide two views of the paleovegetation at any given time, with paleosols providing a broad-scale record of vegetation across the landscape, and mammalian teeth providing a fine-scale record as experienced by the herbivorous mammals who could move about the landscape and feed from specific patches of vegetation.

Finally, microwear analyses suggested that *Sivapithecus* was similar to modern frugivorous great apes and that many contemporaneous species were as frugivorous as their modern counterparts living in rainforests. Isotopic evidence, however, indicates a vegetation mosaic of forest, woodland, and open habitat with little to no extensive closed canopy rainforest. Furthermore, the precipitation regime inferred from intra-tooth equid analyses is indicative of monsoon forests today and habitat more seasonal than modern great apes experience. Given that *Sivapithecus* had a protracted reproductive schedule and a preference for high quality foods, and that it showed no locomotor adaptations that would have allowed it to travel more efficiently in search of food, it is suggested that Siwalik forests were either extensively associated with rivers or differed from modern counterparts in terms of the spatio-temporal availability of fruit. Several scenarios are presented to explain how this difference may have arisen, including a greater diversity of fruiting species in the Miocene, and higher atmospheric CO_2 levels which would have increased plant productivity, followed by a decrease in CO_2 resulting in both the loss of habitat for frugivores and the expansion of C_4 grasses.

Significance

This investigation provides a more detailed analysis than previously recorded of Pakistan Siwalik Miocene habitats for two time levels, and for the preferred habitat of *Sivapithecus*. It is the first detailed study of seasonality over evolutionary time, providing the first paleoprecipitation seasonal regime reconstructions, and it adds to previous estimates of the inception and strengthening of the Asian monsoon. A new method of microwear analysis provides greater detail of fossil diets than previously available. Both the microwear and isotopic data provide the first dietary evidence for some important mammalian species. Isotopic and microwear data from a range of fauna, including equid intra-tooth isotopic sampling, provide evidence for when browsing species first incorporated graze into their diets and when C_4 grass patches first became extensive. This study has also yielded the oldest evidence of C_4 graze known so far. Several hypotheses concerning why C_4 plants gained dominance in the Late Miocene have been proposed (Cerling et al. 1993, 1997b; Quade and Cerling 1995; Pagani et al. 1999), including increasing seasonality, and this investigation provides evidence required to test some of these hypotheses as well. Finally, the intra-tooth isotopic study provides the first evidence for seasonal breeding recorded in fossil teeth.

2 Dental Microwear Analyses of *Sivapithecus* and Contemporaneous Large Mammalian Fauna

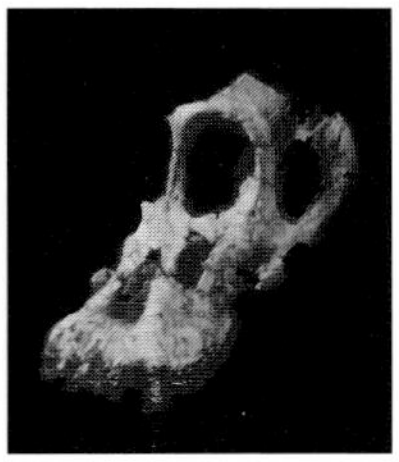

Introduction

Modern frugivorous apes are restricted to habitats where fruit is available almost year-round, with lean periods lasting no more than four months. While orang-utans are an exception, for they are able to tolerate a masting environment by utilizing fat storages, they too are limited to rainforests with high fruit availability. The purpose of this investigation is to determine whether *Sivapithecus* had a diet similar to large fruit-eating apes today, i.e. was heavily dependent upon fruit, and whether its habitat was similar to those of modern apes in terms of fruit availability.

While modern forests in India do not support apes, fossil woods and pollen in India and Nepal indicate that tropical evergreen forests flourished in the Middle Miocene and thus may have been similar to modern ape habitats. These forests were slowly replaced by moist deciduous forest and eventually dry deciduous forest in the Late Miocene (Corvinus and Rimal 2001; Prasad 1993). There is no pollen record found in the Pakistan Siwaliks by which to reconstruct habitat. Fruit availability is therefore best estimated here by the presence of frugivores among the fauna contemporaneous with *Sivapithecus*, with frugivory inferred from dental microwear analyses. Evidence from Gabon and Borneo suggests that many rainforest bovids greatly overlap in fruit choice with primates, sharing 50 percent of the fruit species that they eat (Emmons et al. 1983; Gautier-Hion et al. 1980). Thus evidence of a significant frugivorous component to the fauna in the Siwaliks can be used as an estimate for availability of hominoid-preferred fruits.

Among modern rainforest faunas, many species, such as duikers, are predominantly frugivorous, with fruit making up roughly 70 percent of their diets, comparable to the percentage seen in chimpanzee diets in the same forests. A survey of 126 mammals inhabiting a Gabon rainforest indicates that 56 species, or 44 percent, is frugivorous (Emmons et al. 1983). Among 11 ruminant species, seven, or 64 percent, have diets comprising 70 percent fruit (Gautier-Hion et al. 1980). The authors suggest that the number of mammalian folivores (11 species, 9 percent of the fauna) is surprisingly low in Gabon, and that species folivorous elsewhere, such as *Colobus* at Kibale, incorporate more fruit in their diets at Gabon. The authors attribute this difference to the much higher diversity of fruiting plant species, suggesting at least a three-fold difference compared to other rainforest sites (Emmons et al 1983). Nonetheless, a comparable number of frugivores is reported for rainforests in Malaysia, although these comprise predominantly primates, squirrels, and civets, with only two of five bovids reported as frugivores (Harrison 1962). South American rainforest's fauna are also predominantly frugivorous, comprising 50 to 80 percent of the faunas (Fleming et al. 1987). Thus a strong frugivore component to the Siwalik fauna would suggest a forest of high spatio-temporal availability of fruits, as in rainforests and hominoid habitats today.

Evidence for frugivory among Siwalik fauna in this analysis is inferred from dental microwear. Dental microwear patterns, such as microscopic pits and scratches on the enamel surface, reflect the physical properties of foods most recently eaten by an animal. Turnover in features can be rapid, in some cases as little as 24 hours (Teaford and Oyen 1989). Microwear analyses have been routinely used in investigations of diet in extinct species. Microwear studies in living species have demonstrated significant differences between folivorous and frugivorous primates, soft-fruit eating and hard-fruit eating primates, browsing and grazing ungulates, and even seasonal dietary variation in sympatric hyraxes (Grine and Kay 1988; Teaford and Walker 1984; Solounias and Hayek 1993; Solounias et al. 1988; Walker et al. 1978).

Within primates, much microwear research has focused on methodological issues that might confound dietary interpretations. Variables of concern include patterns resulting from different molar position and facet types (Gordon 1982). Chewing takes place in two phases, as defined by Kay (1977) and Kay and Hiiemae (1974). In Phase I, the mandible moves toward the maxilla until occlusion takes place, and cusps interlock, during which forces are exerted parallel to the occlusal surface, and food is sheared. In Phase II, the mandible moves out of occlusion and forces are applied both parallel and perpendicular to the direction of jaw movement, resulting in both shearing and crushing of food. A total of 11 facets have been defined for primate teeth, eight of which are Phase I and three of which are Phase II (Kay 1977).

Gordon (1982) analyzed differences between Phase I and Phase II facets as well as those due to position in the molar tooth row among eight chimpanzees. She found that while the microwear patterns showed a high degree of regularity, different facet type and molar position both showed consistent differences in microwear, such as feature ratios, densities, and dimensions. Her results suggest that intraspecific differences in microwear may be caused by masticatory biomechanics, and care should therefore be taken in selecting surfaces for interspecific comparisons in order to obtain differences resulting from diet rather than biomechanics. However, Teaford and Walker (1984), in analyzing a number of facets on upper and lower second molars among monkeys and apes, showed that the differences between facets from the same species are generally much less than those between homologous facets from different species. These results suggest that interspecific comparisons on the same facet can accurately depict differences in diet. Furthermore, they demonstrated that homologous mandibular and maxillary facets show the same microwear because upper and lower facets break down food by reciprocal action.

Microwear analyses for Miocene hominoids have generally compared them to modern apes and monkeys (summarized in King 2001), attributing general frugivore/folivore/hard-object diets to them. In general, primate frugivores have higher proportions of pits relative to scratches; folivores, more scratches than pits; with hard-object feeders having the most pits (Teaford and Walker 1984). In a study of modern hominoids, orangutans, which eat fruits with hard shells, showed the highest number of pits, more folivorous gorillas showed the lowest number of pits and the highest number of scratches, and chimpanzees, which eat soft fruits and presumably more fruits than the gorillas of the study, were intermediate for both pits and scratches (Teaford and Walker 1984). This same study included six *Sivapithecus* teeth, and *Sivapithecus* was found to be most similar to chimpanzees.

A study of *Griphopithecus*, a 15 Ma hominoid from the Turkish site Pasalar with similar occlusal morphology to *Sivapithecus*, compared it to orangutans, chimpanzees, and lowland gorillas, and found it most similar to orangutans (King et al. 1999). Again, chimpanzees were intermediate to orangutans and lowland gorillas in many respects, although there were no statistically significant differences between chimps and gorillas. The similarity of chimps and lowland gorillas is not surprising given both ingest substantial amounts of soft fruit (Goodall 1986; Tutin et al. 1991). *Griphopithecus* had higher percentages of pits relative to all the extant hominoids, suggesting a diet similar to orangutans, but with more hard fruits or other hard foods.

Ungar (1996) analyzed microwear in a number of Miocene catarrhines and inferred a variety of diets. A Greek hominoid *Ouranopithecus*, like *Griphopithecus*, appears to have fed on hard objects. *Oreopithecus*, from Italy, resembles modern folivores. All *Dryopithecus* species, as well as *Anapithecus* and *Pliopithecus* species from Rudabanya, appear to have eaten primarily soft fruits.

These previous microwear analyses have provided valuable information on Miocene hominoid diets, but dietary inferences have been limited to general categorizations. This Siwalik investigation, on the other hand, seeks to determine relative proportions of different food categories in a species diet, such as how much of a diet was composed of fruit. In doing so, changes in proportions can be tracked per species over time, allowing inferences of changes in fruit availability. A study at this level of detail requires larger sample sizes than analyzed in previous microwear studies. Previous analyses have been performed using scanning electron microscopes (reviewed by King 2001) and are therefore expensive and require a great deal of time per tooth analyzed. Solounias and Semprebon (2002), however, have developed a new technique using a light microscope to build a library of modern ungulate microwear. This technique has the advantages of being cheaper, simpler, and quicker than using an electron microscope, thus allowing analysis of very large sample sizes. Furthermore, because the new technique involves analyzing larger surface areas, it obtains a better representation of an entire facet rather than a fraction of it. Using this new methodology, Solounias and Semprebon (2002), accurately distinguished between modern browsers, grazers, and mixed feeders. Furthermore, frugivorous bovids, tragulids, and pigs were characterized by large numbers of pits and coarse scratches, analogous to those seen among catarrhines.

This Siwalik investigation uses the new light microscope technique to sample a wide range of species, with many species represented by 20 to 40 specimens spanning millions of years. It then uses a Bayesian statistical approach to analyze results such that diets assigned to fossils are reported in terms of how similar they are to multiple modern species or diet categories. What proportion each food type contributed to the fossil diet is then inferred from these fossil-multiple modern diet comparisons.

Materials and Methods

The Sample

A total of 195 fossil teeth from 13 taxa were analyzed for dental microwear (see Appendix 1 and Table 2.1). All teeth sampled were molars. Taxa represented include *Sivapithecus* spp.; the bovids "2-Keel," *Dorcadoxa porrecticornis*, *Elachistoceras khauristanensis*, *Tragoceridus* spp., and a group of indeterminate small bovids; the tragulids *Dorcabune nagrii*, *Dorcatherium majus*, *Dorcatherium nagrii*, *Dorcatherium* "Sp. D.," and *Dorcatherium* "Y270;" and the suids *Hippopotamodon sivalense* and *Propotamochoerus hysudricus*. These taxa represent all the hominoids, bovids, tragulids, and suids available for analysis that were relatively unweathered. They were chosen because each of these groups contains modern taxa, which include frugivores, and the purpose of this investigation is to estimate fruit availability. All taxa except "2-Keel" were present prior to the extinction of *Sivapithecus*, and all but *Dorcabune nagrii* and *Dorcatherium* "Y270" were present afterwards as well.

Microwear patterns of the fossil teeth were compared to microwear patterns from modern taxa with known diets, including four hominoid taxa (39 specimens), three suid species (29 specimens), 12 browsing ungulate species (195 specimens), nine grazing ungulate species (159 specimens), 16 browse/C_4 graze mix-feed ungulates (259 specimens), five fruit/browse feeders (58 specimens), and one C_3 grazer (nine specimens) (see Table 2.1). All fossil and hominoid microwear data was collected by Nelson. Modern suid data was collected by Nelson and by Solounias and Semprebon (2002). All modern ruminant microwear data was collected by Solounias and Semprebon (2002).

Microwear Analysis

For hominoids, microwear was measured on facet nine, which is a large area located on the lingual surface of the hypoconid on lower molars and on the distobuccal surface of the protocone on upper molars (Kay 1977; Kay and Hiiemae 1974). Upper M2s were selected for modern hominoids, but given the small fossil sample size available, all molars were used for *Sivapithecus*. For modern suids, a medial region of upper M2s was selected. For modern bovids the medial band of the paracone on upper M2s was analyzed. Corresponding regions were analyzed in fossil suids and bovids, but again, given the small sample size available, all molar positions were used.

Microwear was measured on clear epoxy casts of the specimens. All specimens were first cleaned with acetone and cotton swabs. Once dry, each tooth was molded twice using a high precision polyvinylsiloxane dental impression material by President Jet® (Coltene/Whaledent). The first mold was discarded as a final cleaning step. The second mold was surrounded by a ribbon of putty to contain the casting material. Casts were made using a clear epoxy resin (Epo-kwick® resin and hardener from Buehler). Once the epoxy was mixed and poured, the molds were placed in a vacuum chamber to remove air bubbles and were then left to dry for two days.

Table 2.1

Taxa analyzed for microwear

Taxon	Number	Diet
***Sivapithecus* spp.**	**20**	?
Gorilla gorilla beringei	2	Folivore
Gorilla gorilla gorilla	14	Frugivore/ Browse
Pan troglodytes	14	Frugivore/ Browse, piths, herbs
Pongo pygmaeus	9	Hard fruits/ Browse, bark
Propotamochoerus hysudricus	**41**	?
Hippopotamodon sivalense	**23**	?
Babyrousa babyrussa	8	Frugivore/ Browse
Hylochoerus meinertzhageni	7	Folivore
Potamochoerus porcus	14	Fruit and roots/ Browse
Tragoceridus spp.	**8**	?
Indeterminate small bovids	**7**	?
Elachistocerus	**2**	?
Dorcadoxa porrecticornis	**5**	?
"2-Keel"	**2**	?
Dorcatherium "sp. D"	**31**	?
Dorcatherium majus	**32**	?
Dorcabune nagrii	**10**	?
Dorcatherium "Y270 sp."	**2**	?
Dorcatherium nagrii	11	?
Alces alces	15	Browse
Antilocapra americana	42	Browse
Boocercus euryceros	15	Browse
Budorcas taxicolor	15	Browse
Camelus bacterianus	3	Browse
Camelus dromedarius	6	Browse
Diceros bicornis	11	Browse
Giraffa camelopardalis	23	Browse
Litocranius walleri	19	Browse
Okapia johnstoni	14	Browse
Tragelaphus imberbis	16	Browse
Tragelaphus strepsiceros	10	Browse
Capreolus capreolus	6	Browse
Alcelaphus buselaphus	6	C4 Graze
Bison bison	18	C4 Graze
Cervus duvauceli	25	C4 Graze
Connochaetes taurinus	27	C4 Graze
Equus burchelli	47	C4 Graze
Equus grevyi	11	C4 Graze
Hippotragus niger	5	C4 Graze
Kobus ellipsiprymnus	12	C4 Graze
Tetracerus quadricornis	8	C4 Graze
Axis axis	39	Mixed Browse/C4
Boselaphus tragocamelus	9	Mixed Browse/C4
Capricornis sumatraensis	9	Mixed Browse/C4
Cervus canadensis	28	Mixed Browse/C4
Gazella granti	29	Mixed Browse/C4
Gazella thomsoni	19	Mixed Browse/C4
Lama glama	6	Mixed Browse/C4
Muntiacus muntjak	10	Mixed Browse/C4
Taurotragus oryx	17	Mixed Browse/C4
Tragelaphus scriptus	31	Mixed Browse/C4
Vicugna vicugna	7	Mixed Browse/C4
Capra hircus	4	Mixed Browse/C4
Capra ibex	7	Mixed Browse/C4
Cervus unicolor	5	Mixed Browse/C4
Ovibos moshatus	20	Mixed Browse/C4
Ovis canadensis	19	Mixed Browse/C4
Cephalophus dorsalis	21	Fruit/Browse
Cephalophus natalensis	4	Fruit/Browse
Cephalophus niger	7	Fruit/Browse
Cephalophus silvicultor	15	Fruit/Browse
Tragulus spp.	11	Fruit/Browse
Equus caballus	9	C3 Graze

Taxa analyzed for microwear representing three datasets—hominoids, suids, and ruminants. Fossil taxa are in bold.

Casts were subsequently analyzed under a light microscope at 35X magnification using the new technique of Solounias and Semprebon (2002). The surface measured was 0.5 mm-x-0.5 mm, and two surfaces were measured per tooth, with averages reported for continuous variables. Characters measured included number of pits and number of scratches; the presence/absence of at least four gouges, at least four large pits, and cross scratches; and the nature of the scratches, i.e. whether they were predominantly coarse, fine, or a mix of coarse and fine. Pits are defined as those features with a length:width ratio below 4:1 and are generally circular, while scratches have ratios greater than 4:1. Gouges are very irregularly shaped large pits. Cross scratches are those running perpendicular to one another.

Data Analysis

Data were analyzed using a Bayesian general location model with diet category for fossil teeth treated as missing data. Bayesian statistics, unlike classical statistics, allows for *a priori* knowledge to be informative in making statistical inferences (McGee 1971; Morgan 1968). Bayesian inference is based upon a quantity called the posterior probability. Prior probabilities (*a priori* knowledge) is converted to posterior probabilities using Bayes Theorem, as follows:

$$p(B|A) = p(B)\,\frac{p(A|B)}{p(A)}$$

Where $p(B|A)$, or the probability of B given A, is the posterior probability, $p(B)$ is the prior probability, and $\frac{p(A|B)}{p(A)}$ is the maximum likelihood.

When prior probabilities are chosen to be noninformative (i.e. $p(B) = 1$), then the posterior probability equals the maximum likelihood estimate.

For example, an investigator is attempting to determine whether one of two hypotheses is true (H_1 and H_2) and obtains some preliminary data. The first question asked is, given prior knowledge, how likely is H_1? This probability becomes the prior probability $p(H_1)$. The next question is, what are the chances of obtaining the data (D) if H_1 is true? This probability has to be compared to the probability of obtaining D no matter what, and the ratio of those probabilities is the likelihood ratio. Note that the likelihood ratio is therefore a conditional probability.

Under this scenario, using Bayes Theorem, the posterior probability can be calculated as follows:

$$p(H_1(D) = p(H_1)\,\frac{p(D(H_1)}{P(D)}$$

The Bayesian approach is useful in this microwear study because it allows model parameters (probabilities of categorical variables such as presence of large pits and means of continuous variables such as number of scratches) and the missing data (diets of fossil taxa) to be treated as random variables with distributions. Thus diets assigned to fossil taxa can be reported in terms of degree of similarity to multiple modern species or diets, allowing for the possibility that there is no perfect modern analogue for a fossil diet. This method allows for microwear characters to represent specific food preferences, and fossil diets can be interpreted in terms of how much each food type contributed to the diet. Furthermore, short-term variability of a fossil species' diet, such as percentages of fruit versus browse, can be inferred from variability between specimens. Finally, while in this study there was no *a priori* knowledge to contribute to diet determination, such information could be built into the model in future studies via prior probabilities.

This microwear analysis incorporates both categorical variables and continuous variables and therefore uses a general location model to estimate likelihood ratios. The general location model is a multivariate correlation model with mixed discrete and continuous variables (Olkin and Tate 1961). The model posits a loglinear model on a contingency table for categorical variables, and within each cell of the contingency table, it posits a multivariate normal distribution for the continuous variables.

Loglinear Model

Categorical data is most easily summarized by a *contingency table*. For data with two variables x and y, the contingency table is simply a two-way table with a row for each level of x and a column for each level of y. Each cell of the table contains the number of observations with those values of x and y. For example, cell i,j of the contingency table is equal to the number of observations for which $x = i$ and $y = j$. Contingency tables for datasets with more than two variables are defined similarly, with the number of dimensions of the table equal to the number of categorical variables. Statistical inference for categorical variables involves estimating the probability that an observation lies in a given cell of the contingency table.

When datasets have small numbers of observations and/or large numbers of variables, there are often *random zeroes* in the contingency table, meaning cells with no observations due to chance. When random zeroes appear in the dataset, the corresponding cell probabilities cannot be estimated consistently, and more structure is needed on the model for the contingency table. Loglinear models supply the needed structure for such datasets. In general, loglinear models allow for different dependence relations among the categorical variables.

One of the simplest loglinear model is the independence model, which allows no associations among variables, i.e. all variables are considered to be independent. Other loglinear models may include all two-way associations between variables, all three-way associations, etc.

Linear Model on Cell Means

In addition to creating a loglinear model on the categorical variables, the general location model posits a multivariate normal model on the continuous variables, with the mean allowed to vary over the cells of the contingency table. As with the categorical variables, if the number of observations is fairly small compared to the number of cells in the contingency table, it may be difficult or impossible to estimate all the parameters associated with the continuous variables. More structure is added to the model by placing a linear model on the means of the continuous variables. Similar to the loglinear model discussed above, this consists of limiting the associations allowed among means of each cell.

Analysis of Microwear Results

Using the general location model, a contingency table was first established for modern species of known diets. Each possible combination of results for all of the variables measured (microwear characters) is represented by a cell in the contingency table, and modern species or diets are mapped onto those cells according to the combination of results observed. Each fossil specimen, via the general location model, is then compared to the contingency table of microwear representing known diets, and an initial posterior probability is assigned to it. The posterior probability involves a summation over all possible outcomes, or combinations of variables, and is nearly impossible to calculate analytically. A numerical method was therefore used to approximate the posterior probability. This process involves sampling the possible outcomes within the posterior probability many times in order to generate the mean of the posterior probability. These iterative samples, known as draws, proceed by Data Augmentation (Tanner and Wong 1987), an example of a Markov Chain Monte Carlo algorithm.

For example, if a *Sivapithecus* molar were compared to a contingency table consisting of chimpanzees and orangutans, the initial probability might be 0.4 chimpanzee, 0.6 orangutan. Probabilities for each fossil tooth always sum to 1.0. Using an analogy, an "unfair coin toss" is then performed 2,000 times. In the *Sivapithecus* example above, the coin is weighted 0.4 on the chimpanzee side and 0.6 on the orangutan side. Probabilities from these draws are then averaged to yield final posterior probabilities for the particular specimen.

In addition to determining initial probabilities for the missing data, the model also determines initial parameters for the variables of that "missing" fossil species, i.e. means for the continuous variables, probabilities for the categorical variables. This means that the "coin toss" is more complicated than described above. The algorithm actually prescribes iteratively drawing values (2,000 times) of the missing data, conditional on the observed data and current parameters, and drawing the parameters, conditional on the observed data and current missing data. Using the coin toss analogy, at the same time that the algorithm is estimating probabilities from these coin tosses, it is also determining parameters of the variables (means and probabilities) based upon these coin tosses. This microwear study, however, is only interested in estimates of the missing data, or assignment of diet to a fossil specimen, once the parameters have been determined. It is not interested in estimates of parameters based upon the missing data. However, after a sufficient number of draws, or sampling of all possible posterior probabilities, the algorithm converges on the mean parameters for the true posterior probability. A conservative "burn-in" period was therefore allowed, meaning the first 1,000 draws, or coin tosses, were discarded, and only the estimates of posterior probabilities of the missing data from the last 1,000 draws were considered. For these draws, estimates of parameters had become fixed.

In this investigation, three datasets were considered- a set of fossil and living hominoids, one of fossil and modern suids, and one of fossil bovids and tragulids and modern ruminants. Data Augmentation converged for each dataset after a conservative 1,000 iteration burn-in period. All prior distributions were chosen to be non-informative (prior probability set to equal one), i.e. *Sivapithecus* was not assumed to have a diet of any living hominoid. Variables were treated as independent of one another, and initial experiments with different model constraints, or associations among variables, had very little effect on the results. Diets of fossil specimens are presented as similarities to modern species in terms of posterior probabilities, and these probabilities always sum to 1.0. Diets for fossil taxa are then estimated from averaged probabilities of individual specimens.

Before inferring diets of fossil species, it was important to determine how well this new microwear method distinguished among the different diets of the modern species. Again, a general location model was used. This time, for each dataset of modern species, one specimen was removed and treated as missing for the species diet variable, as if it were a fossil. After Data Augmentation, as with the fossil specimens, posterior probabilities were produced from a contingency table. For comparison, a baseline "null model" which did not distinguish living taxa was created by repeating the above process many times, each time randomly permuting the species/diet variable before the analysis. This allowed for determination of how often the model predicts the correct values for the species/diet variable when that variable is completely arbitrary, that is, when a diet has been randomly assigned to each modern specimen of the contingency table. A p-value was then generated which is the proportion of times the model using the random permutations predicted the values better on average than did the model with the correct group indicators.

Results and Discussion

All microwear measurements for each fossil tooth are listed in Appendix 1.

Hominoid Dataset

Facet nine on all upper and lower molars were analyzed for *Sivapithecus* because sample size would have been too small if only M2s were used. However, Kruskal-Wallis One-Way Anova analyses for number of pits and number of scratches indicated no significant differences between M1s, M2s, and M3s, suggesting results were not affected by molar position.

Mountain gorilla microwear analyses yielded a very high number of mixed (fine and coarse) scratches compared to other hominoids, illustrating a diet of more fiber, and a low number of pits, indicative of less fruit-eating (see Appendix 1). Orangutan microwear was characterized by a high number of pits and mixed scratches, indicative of feeding on hard fruits. Chimpanzee microwear included a high number of pits (from fruits) and a low number of scratches, with many of them fine, suggesting fewer hard objects in the diet compared to orangutans. Lowland gorillas showed a relatively low number of pits on average and an intermediate number of mixed scratches, indicating less fruit than chimpanzees or orangutans but a less fibrous diet than that of mountain gorillas.

As judged by the general location model, the new microwear method was able to distinguish between the diets of large living apes well, especially given the overlap in their frugivorous diets, excepting the mountain gorilla. A p-value of 0.005 was generated when the following question was asked: How often was a permuted, arbitrary dataset better able to predict diet for each and every living species? A p-value of 0.09 was generated when the question was changed to: How often did the permuted dataset better determine diet for the living taxa on average?

Sivapithecus microwear on average resembles that of orangutans and chimpanzees in having a high number of pits. Its intermediate number of mixed scratches resembles orangutan microwear more than chimpanzee's. The posterior probabilities per tooth for *Sivapithecus* with respect to each living hominoid are listed in Table 2.2. On average, the posterior probabilities for *Sivapithecus* are as follows: mountain gorilla (0.08), lowland gorilla (0.27), chimpanzee (0.30), and orangutan (0.35). *Sivapithecus* therefore looks most similar to an orangutan and unlike a mountain gorilla, but not unlike a lowland gorilla or chimpanzee (see Figure 2.1). On an individual specimen level, five individuals give a dietary signal that is predominantly chimpanzee (a probability greater than 0.5); six, predominantly orangutan; one, predominantly lowland gorilla; and eight, a mix of all three frugivores. These results indicate that *Sivapithecus* was primarily frugivorous, as are most modern great apes, and its diet contained some hard objects, like that of an orangutan. However, *Sivapithecus* does not greatly resemble any particular extant frugivorous hominoid.

Two *Sivapithecus* teeth (Y4230 and Y4622) were characterized by a very high number of pits. *Sivapithecus* teeth on average had 30.2 pits, while these teeth had pit counts of 53.5 and 58.5 (see Appendix 1). The microwear for both of these teeth looks predominantly like that of an orangutan (orangutan probabilities of 0.74 and 0.67), and both come from locality Y182. No other *Sivapithecus* teeth were sampled from this site. These two teeth suggest that locality Y182 might have been characterized by some vegetation which produced hard nuts or hard-covered fruits or seeds, like those of the dipterocarps that orangutans eat today (MacKinnon 1974; Rodman 1977).

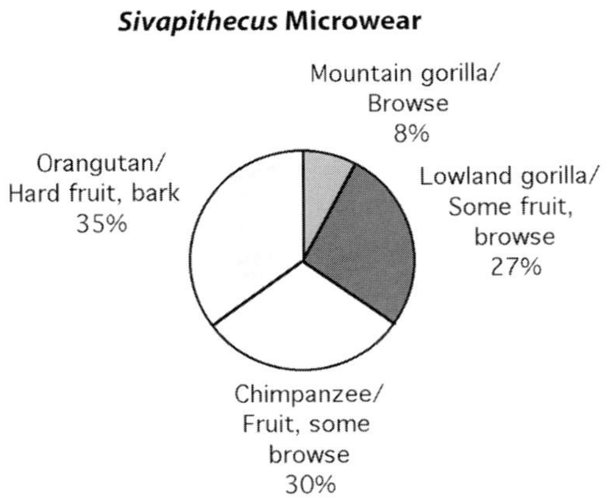

Figure 2.1 Microwear results for *Sivapithecus* indicate that its diet resembled that of modern frugivorous hominoids and included some hard objects, such as hard-covered fruits or seeds.

If numbers of pits and scratches are considered irrespective of the categorical variables, then the ratio of pits to scratches (reported as pits/scratches) also suggests that *Sivapithecus* is most similar to orangutans (see Table 2.3 and Figure 2.2). Among the living hominoids, mountain gorillas showed the lowest ratio (0.60) (due to a low number of pits and high number of scratches), followed by lowland gorillas (1.1) and orangutans (1.5), with chimpanzees having the highest ratio (2.1) (due to a low number of scratches). *Sivapithecus* (1.6) falls between orangutans and chimpanzees, but very close to orangutans. The order of living taxa with respect to the pit:scratch ratio is the same as that found by King et al. (1999) using a scanning electron microscope, although the ratios are higher using the light microscope (possibly because of the high refractivity of tiny pits under the light microscope making them more visible). King found that *Griphopithecus alpani* from Pasalar had the highest ratio of all and the highest number of pits, suggesting it fed on harder items than either living hominoids or *Sivapithecus*. These results differ from those of Teaford and Walker (1984), who found that orangutans had the highest ratio, and *Sivapithecus* was most similar to chimpanzees in their study. The discrepancy between King et al. (1999) and Teaford and Walker (1984) could be due to differences in sample size, with a greater number of specimens analyzed in King et al. (1999). Likewise, the Siwalik analysis included 20 *Sivapithecus* specimens, while Teaford and Walker (1984) analyzed only six.

A Kruskal-Wallis One-Way ANOVA indicates that differences in number of pits is marginally significant between all hominoid taxa (*Chi*-Square = 8.79; p = 0.067), and differences in number of scratches is significant (*Chi*-Square = 18.03; p = 0.001). Using Mann-Whitney analyses corrected for ties, Sivapithecus is indistinguishable from orangutans but differs from all other taxa in either number of pits or scratches in the same pattern that orangutans differ from all other extant hominoids (see Tables 2.4 and 2.5). Chimpanzees have fewer scratches than all other hominoids. Lowland gorillas have the fewest pits except for mountain gorillas (which were only represented by two specimens). Mountain gorillas have more scratches than all other hominoids.

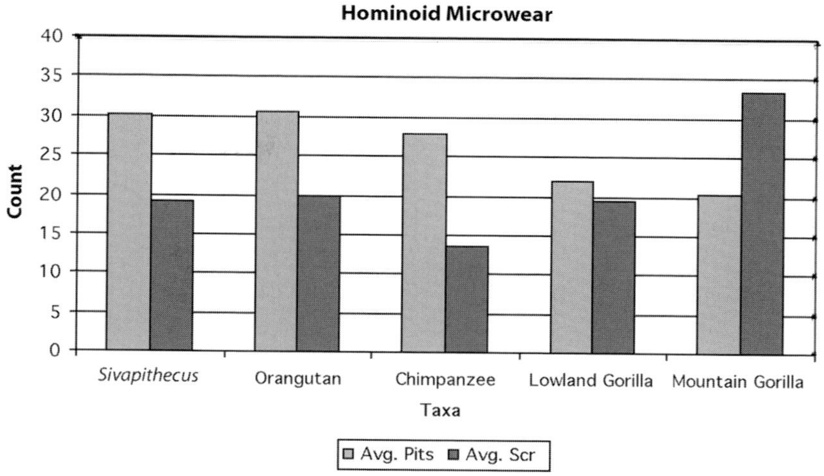

Figure 2.2 Average pit and scratch counts for each hominoid taxon. *Sivapithecus* is indistinguishable from orangutans.

Suid Dataset

Two Siwalik suids, *Propotamochoerus hysudricus* (50 kg) and the very large *Hippopotamodon sivalense* 0 kg) were compared to three modern suids, *Babyrousa babyrussa*, *Hylochoerus meinertzhageni*, and *Potamochoerus porcus*. *Potamochoerus*, the African red river hog, is frugivorous and rootles as well. *Babyrousa*, the pig deer from southeast Asia, feeds on fallen fruit but does not rootle. *Hylochoerus*, the giant forest hog of Africa, is folivorous, feeding primarily on short grasses, sedges, shrubs, and herbaceous growth, with very little digging or grubbing and little to no fruit (Kingdon 1997; Nowak 1999).

Potamochoerus and *Babyrousa*, the two fruit eaters, are distinguished by a large number of pits, including large pits, and coarse and mixed scratches. *Potamochoerus* has a higher number of pits on average (34.8, n = 14) than *Babyrousa* (26.4, n = 8), most likely due to both fruit and root eating. *Hylochoerus*, on the other hand, had fewer pits, few large pits, and predominantly fine scratches, indicative of more fibrous foods as opposed to fruit. Analysis with the general location model indicates that data obtained using the light microscope technique amply differentiated between the diets of the living suids. The p-value generated when asking whether the permuted dataset distinguished better for all of the living taxa was 0.003; that for asking whether it distinguished better the living taxa on average was 0.04.

Both fossil suids, *Propotamochoerus* and *Hippopotamodon*, have microwear features similar to the extant fruit-eating pigs—a high number of pits, including large pits, and predominantly mixed scratches. The average posterior probabilities for *Propotamochoerus* suggest that its diet is most similar to that of *Potamochoerus* (0.44), followed by *Hylochoerus* (0.36), and is least like *Babyrousa* (0.21) (see Table 2.6 and Figure 2.3). These results indicate that *Propotamochoerus* was omnivorous but predominantly frugivorous and probably rootled as well. However, when considered individually, some specimens yield very little to no fruit signal. If the percentage fruit/foliage in the diet is directly inferred from the posterior probabilities, using a 65 percent or greater criterion for frugivory (comparable to percentages seen in frugivores from Gabon [Emmons et al. 1983; Gautier-Hion et al. 1980]) and likewise for folivory, 24 individuals give a primarily frugivorous dietary signal, seven are most like the folivorous *Hylochoerus*, and 10 give a fruit/foliage combination dietary signal. These results suggest that while fruit was the primary component of *Propotamochoerus*'s diet, this fruit may have only been seasonally abundant, with significant periods of little fruit availability during which *Propotamochoerus* became folivorous.

Table 2.2

Posterior probabilities for *Sivapithecus*

Specimen ID	Mountain Gorilla	Lowland Gorilla	Chimpanzee	Orangutan	Age (Ma)	Locality
Y9986	0.03	0.40	0.20	0.37	10	Y311
Y11533	0.01	0.24	0.63	0.12	10	Y311
Y11998	0.25	0.28	0.05	0.42	10	Y311
Y11999	0.01	0.16	0.76	0.07	10	Y311
Y5001	0.04	0.46	0.40	0.09	9.3	Y207
Y13566	0.06	0.39	0.23	0.32	9.3	Y350
Y15000	0.08	0.20	0.06	0.66	9.3	Y410
Y15030	0.05	0.40	0.26	0.29	9.3	Y416
Y4230	0.11	0.09	0.06	0.74	9.2	Y182
Y4622	0.25	0.07	0.01	0.67	9.2	Y182
Y5260	0.02	0.27	0.19	0.52	9.2	Y211
Y9563	0.00	0.13	0.77	0.10	9.2	Y260
Y9899	0.05	0.41	0.14	0.40	9.2	Y260
Y9972	0.01	0.10	0.58	0.31	9.2	Y260
Y9977	0.11	0.10	0.05	0.74	9.2	Y260
Y13165	0.13	0.11	0.05	0.71	9.2	Y260
Y11704	0.07	0.35	0.32	0.26	9.2	Y317
Y11707	0.29	0.57	0.08	0.07	9.2	Y317
Y11708	0.03	0.36	0.46	0.15	9.2	Y317
Y16077	0.01	0.26	0.66	0.07		Y498
AVERAGE	**0.08**	**0.27**	**0.30**	**0.35**		
STDEV	**0.09**	**0.15**	**0.26**	**0.24**		

Posterior probabilities for *Sivapithecus* teeth. *Sivapithecus* looks most similar to orangutans but resembles frugivorous hominoids in general. Posterior probabilities across extant species categories sum to 1.0. $N = 20$.

Table 2.3

Average pits and scratches counts for hominoids

	Pit Count	Scratch Count	Pits/Scratches
Mountain Gorilla (2)	20.3	33.5	0.61
Lowland Gorilla (14)	21.9	19.5	1.12
Orangutan (9)	30.7	19.9	1.54
Chimpanzee (14)	27.9	13.6	2.05
Sivapithecus spp. (20)	30.5	19.0	1.61

Average counts of pits and scratches per tooth for hominoid taxa. *Sivapithecus* most resembles orangutans in counts and in pit: scratch ratio. Number of specimens in parentheses.

The average probabilities for *Hippopotamodon* suggest that its diet was most similar to *Potamochoerus* as well (0.49), followed by *Hylochoerus* (0.28) and *Babyrousa* (0.24) (see Table 2.7 and Figure 2.3). Again, this suid appears to have been frugivorous with some rootling. Considered individually, using the same criterion mentioned above, 14 specimens look frugivorous, one looks folivorous, and eight suggest a mixed diet. These results suggest a diet that contained at least some fruit even during lean periods, unlike that of *Propotamochoerus*, though it should be noted that the sample size is smaller for *Hippopotamodon* (23 compared to 41).

When individuals for each species are divided into those that lived prior to the extinction of *Sivapithecus* (8.4 Ma or older), and those that were present afterwards, no significant differences are found in numbers of pits or scratches or in posterior probabilities over time for either taxon, suggesting no change in diet over time. Three *Propotamochoerus* teeth have a very high number of pits relative to the mean for this taxon (58.5 for Y13678, 55 for Y4554, and 47 for Y7654 compared to the average of 30.7; see Appendix 1). Like the *Sivapithecus* specimens, two of these specimens come from locality Y182, again suggesting hard-covered fruits. If from dipterocarps, this result would also explain the seasonal variability in fruit availability suggested by the *Propotamochoerus* teeth. *Hippopotamodon* was represented by one tooth from locality Y182, and it shows the average number of pits for this taxon (about 28). *Hippopotamodon* may have been feeding on fruits different than those of *Propotamochoerus* and as a result may have been less susceptible to seasonal changes. Alternatively, locality 182 probably represents a time period of thousands of years, and these suids were not contemporaneous.

Propotamochoerus hysudricus Microwear

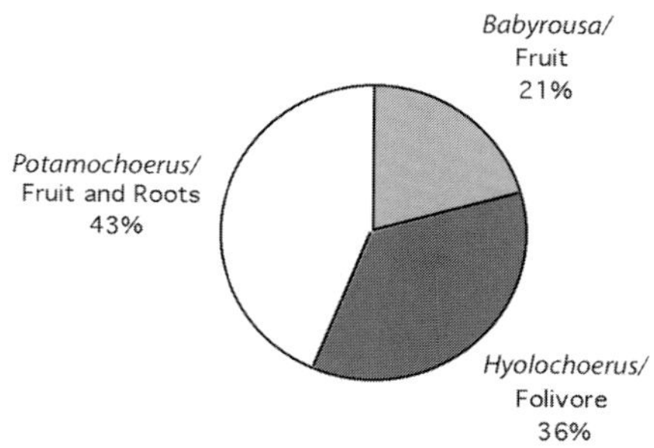

Hippopotamodon sivalense Microwear

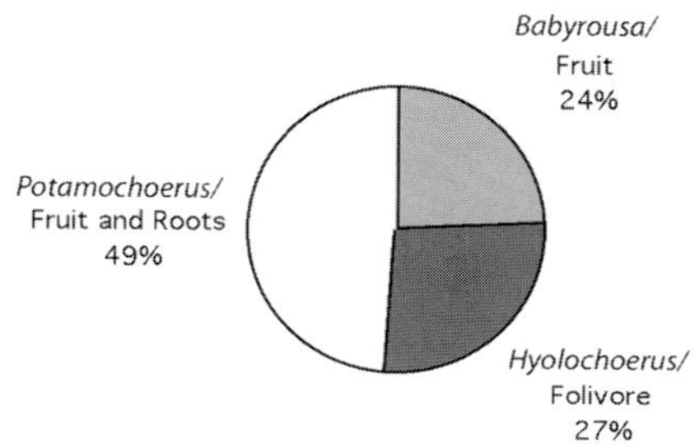

Figure 2.3. Microwear results for *Propotamochoerus hysudricus* ($n = 41$) indicate a diet of fruit and roots with periods of no fruit availability. *Hippopotamodon* ($n = 23$) also fed on fruit and roots.

Bovid and Tragulid Dataset

Five fossil tragulid species, four fossil bovid genera, and one group of indeterminate small fossil bovids were compared to 43 species in the following categories of ruminants: browser, C_4 grazer, C_3 grazer, mixed feeder, and frugivore/ browser. Note that unlike the previous datasets, fossils are compared to diets represented by numerous species, rather than to individual species. The mixed feeder category specifies species which eat significant amounts of both browse and C_4 graze; it does not include species that eat a considerable amount of fruit.

Defining features of C_4 grazers based on Solounias and Semprebon (2002) are high numbers of scratches with almost all scratches being coarse. Browsers, by contrast, generally have low numbers of scratches, with many of them being fine. C_3 grazers are characterized by high numbers of fine scratches, possibly due to smaller phytoliths in C_3 grasses versus C_4 grasses (Solounias, personal communication). Mixed feeders cluster between browsers and grazers for average number and texture of scratches, and scratch numbers found among individuals within a species often show a bimodal distribution if they are seasonal mixed feeders (Solounias and Semprebon in 2002).

The frugivore/browser dietary category is the most difficult to distinguish among living taxa, for the microwear generated by feeding on fruit is comparable to that of mixed feeders and grazers. Frugivorous duikers and tragulids included in the original study by Solounias and Semprebon (2002) fell within the mixed feeding and grazing categories due to high numbers of coarse scratches. These coarsely textured scratches are most likely the result of feeding on hard-covered fruits or seeds which may scratch teeth in a way similar to phytoliths in C_4 grasses. These hard fruits leave higher numbers of pits than C_4 graze, however, allowing for some distinction between the different diets of heavy abrasion. A better way to distinguish frugivores from mixed or grazing diets is to incorporate both microwear and isotopic analyses in determination of unknown diets. C_3 and C_4 plants differ in ratios of carbon stable isotopes, and isotopic analyses of tooth enamel can detect C_4 graze in the diet. Many taxa analyzed in this microwear study were also analyzed isotopically, as reported in further detail in Chapter Three.

Table 2.4

Mann-Whitney comparisons of pit counts for hominoids

	Sivapithecus	Orangutan	Chimpanzee	Lowland Gorilla	Mountain Gorilla
Sivapithecus		NS	NS	*Z*=-2.45; *p*=0.014	NS
Orangutan			NS	*Z*=-2.05; *p*=0.04	NS
Chimpanzee				*Z*=-2.00; p=0.045	NS
Lowland Gorilla					NS

Mann-Whitney pair-wise comparisons for pit counts among hominoids.

Table 2.5

Mann-Whitney comparisons of scratch counts for hominoids

	Sivapithecus	Orangutan	Chimpanzee	Lowland Gorilla	Mountain Gorilla
Sivapithecus		NS	*Z*=-3.19; *p*=0.001	NS	*Z*=-2.06; *p*=0.04
Orangutan			*Z*=-2.87; *p*=0.004	NS	*Z*=-1.89; *p*=0.059
Chimpanzee				*Z*=-2.92; *p*=0.004	*Z*=-2.23; *p*=0.023
Lowland Gorilla					*Z*=-2.07; *p*=0.038

Mann-Whitney pair-wise comparisons for scratch counts among hominoids.

Results from the general location model analysis indicate that the new microwear technique distinguished very well between modern dietary categories, yielding a *p*-value < 0.01 when it was asked whether a permuted dataset distinguished better for any of the groups. This success is most likely due to the very large sample size (43 species, 680 specimens) (see Table 2.1).

As with modern tragulids, the wear on each fossil tragulid species analyzed in this study spans all dietary categories except C_3 graze as shown by the large numbers of pits and coarse scratches (see Figure 2.4 and Tables 2.8 through 2.12). Isotopic analyses for *Dorcabune nagrii* and *Dorcatherium majus* indicate no C_4 in the diet, and analyses of a wide range of taxa indicate that only horses incorporated any C_4 in their diets prior to 8.0 Ma (see Chapter Three). Therefore, the heavy wear among the tragulids is most likely due to feeding on hard-covered fruits and seeds, and probabilities in the graze and mix feeder categories most likely represent coarse wear resulting from fruit rather than phytoliths. The best determinant of amount of fruit versus non-fruit in the diet is therefore the browsing category, for it comprises wear features not associated with fruit. Tragulids with the smallest browse component, and hence the largest fruit component, are the three large species, *Dorcabune nagrii*, *Dorcatherium majus*, and *Dorcatherium* "Sp. D." (see Figure 2.4 and Tables 2.8 to 2.10). In each of these species, the average posterior probability for the browse category is less than 0.30. If percentage of the diet is inferred from probabilities, then this result suggests that the diets of these tragulids comprised at least 70 percent fruit, comparable to frugivorous ruminants in Gabon today (Emmons et al. 1983; Gautier-Hion 1980). At the individual level, six of ten *Dorcabune nagrii* specimens, or 60 percent, have probabilities suggesting a diet of less than 35 percent browse. None have diets that are over 65 percent browse. For both *D. majus* and *Dorcatherium* "Sp. D." over 70 percent of specimens (n = 32 and 31, respectively) indicate a diet comprised of less than 35 percent browse. For both species, however, there are two specimens that indicate diets of over 65 percent browse, suggesting there may have been periods of fruit shortages.

Dorcatherium "Y270 sp." is a small tragulid represented by only two teeth, but both of those suggest a diet of roughly 30 percent browse (see Figure 2.4 and Table 2.11). *Dorcatherium nagrii* is another small species, and its microwear suggests a stronger browse

Table 2.6

Posterior probabilities for *Propotamochoerus hysudricus*

Specimen ID	*Babyrousa*	*Hylochoerus*	*Potamochoerus*	Age (Ma)	Locality
Y22080	0.97	0.03	0.01	10	Y311
Y17578	0.19	0.30	0.51	10	Y311
Y17088	0.07	0.35	0.58	9.5	Y262
Y14044	0.66	0.01	0.33	9.5	Y262
Y6226	0.13	0.54	0.33	9.5	Y228
Y13587	0.04	0.81	0.16	9.3	Y227
Y6011	0.06	0.64	0.30	9.3	Y341
Y41368	0.08	0.68	0.24	9.3	Y269
Y10224	0.09	0.46	0.45	9.3	Y309
Y6137	0.13	0.59	0.29	9.3	Y221
Y14501	0.17	0.47	0.36	9.3	Y269
Y14502	0.34	0.05	0.61	9.3	Y269
Y51848	0.38	0.05	0.57	9.3	Y269
Y16132	0.39	0.06	0.55	9.3	Y310
Y5112	0.01	0.95	0.04	9.2	Y211
Y7654	0.08	0.29	0.63	9.2	Y182
Y4554	0.18	0.02	0.81	9.2	Y182
Y11007	0.17	0.43	0.41	9.2	Y314
Y9355	0.19	0.27	0.54	9.2	Y260
Y10998	0.20	0.38	0.43	9.2	Y314
Y4599	0.18	0.24	0.59	8.9	Y193
Y13678	0.16	0.06	0.78	8.9	Y367
Y17968	0.10	0.59	0.30	8.5	Y604
Y4813	0.17	0.23	0.60	8.5	Y203
Y457	0.52	0.07	0.41	8.1	Y24
Y49515	0.12	0.09	0.79	8	Y599
Y49840	0.04	0.65	0.31	7.8	Y906
Y49844	0.27	0.27	0.46	7.8	Y906
Y50880	0.35	0.02	0.64	7.8	Y949
Y18124	0.24	0.17	0.59	7.2	Y452
Y14045	0.07	0.77	0.16	7.1	Y370
Y49816	0.04	0.87	0.09	7	Y916
Y52855	0.01	0.97	0.02		L084
D3	0.05	0.76	0.20		D005
Y19698	0.27	0.05	0.69		
Y51575	0.17	0.44	0.40		Y539
Y2807	0.06	0.19	0.76		Y105
Y52623	0.18	0.26	0.56		Y960

Table 2.6 Continued

Posterior probabilities for *Propotamochoerus hysudricus*

Specimen ID	*Babyrousa*	*Hylochoerus*	*Potamochoerus*	Age (Ma)	Locality
Y19803	0.21	0.41	0.38		
Y85	0.24	0.12	0.63		Y010
Y19748	0.48	0.01	0.50		L37
AVERAGE	**0.21**	**0.36**	**0.44**		
STDEV	**0.19**	**0.29**	**0.21**		
Pre 8.4 Ma Avg	**0.21**	**0.35**	**0.43**		
Post 8.4 MaAvg	**0.20**	**0.37**	**0.43**		

Posterior probabilities for *Propotamochoerus hysudricus* microwear. *Propotamochoerus* most resembles frugivores with some rootling, but some specimens indicate periods of little fruit availability. Ages and localities listed when available. No differences are found between individuals living prior to *Sivapithecus*'s extinction vs. those that lived afterwards. $N = 41$.

component, with an average posterior probability of 0.43 for the browse category (see Figure 2.4 and Table 2.12). At the individual level, eight of 11 specimens, or 73 percent, have a browse component over 35 percent, and three of the 11, or 27 percent, have diets of over 65 percent browse. These results suggest that fruit was a smaller component of *Dorcatherium nagrii*'s diet relative to modern tragulids. *Dorcatherium nagrii* was no larger than modern tragulids, weighing less than 15 kg. Modern ruminants of this size today, given their high metabolic requirements, require a non-fibrous diet of high protein content, such as fruit or very selective browse including buds and young leaves (Janis 1982, 1990). Many modern small antelopes and duikers eat primarily fresh buds, leaves, and shoots, but they are restricted to moist forests where plants produce new foliage year-round (Kingdon 1997). A comparison to modern analogues therefore suggests that *Dorcatherium nagrii* was a highly selective browser and must have inhabited moist forest with year-round production of new foliage. Differences in posterior probabilities among individual specimens for this species could reflect seasonal differences in diet, with fruit making a greater contribution during certain seasons, or they could reflect geographic differences in fruit availability as well.

When individuals from all tragulid species are combined and then divided into those that lived prior to the extinction of *Sivapithecus* (65 individuals) and those that were present afterwards (16 individuals), the posterior probability for the browsing dietary category significantly increases from 0.26 to 0.37 over time (Mann-Whitney, corrected for ties, $Z = -2.03$, $p = 0.043$), suggesting that fruit is becoming a smaller component of tragulid diets in general (see Figure 2.5). For *Dorcatherium nagrii* (the tragulid with the largest browse component), the proportion of browse in the diet, inferred from browse posterior probability, increases from 42 percent prior to the extinction of *Sivapithecus* ($n = 7$), to 61 percent afterwards ($n = 3$), but this increase is not statistically significant (see Table 2.12). *D. majus* and *Dorcatherium* "Sp. D." also show minor increases in browse, but again these are not significant (see Tables 2.9 and 2.10). *Dorcatherium* "Sp. D." does show significant decreases in numbers of pits (from 36.5 in 24 older specimens to 25.5 in seven younger) and scratches (from 16.2 to 12.4) (pits: $Z = -2.67$; $p = .008$; scratches: $Z = -2.30$, $p = 0.022$), which suggests a shift to feeding less on hard objects. Finally, *Dorcabune nagrii* and *Dorcatherium* "Y270" become extinct around the same time as *Sivapithecus* or earlier. *Dorcabune nagrii* disappears around 8.5–8.4 Ma, while the youngest specimen of *Dorcatherium* "Y270" dates to 9.2 Ma, with an inferred last appearance for this species at 8.6 Ma (Barry et al. 2002).

Among the fossil bovid taxa analyzed, *Tragoceridus* spp. and *Dorcadoxa porrecticornis* are medium-size (about 80 kg), "2-Keel" is eland-size (400–1,000 kg), *Elachistoceras khauristanensis* is less than 15 kg, and

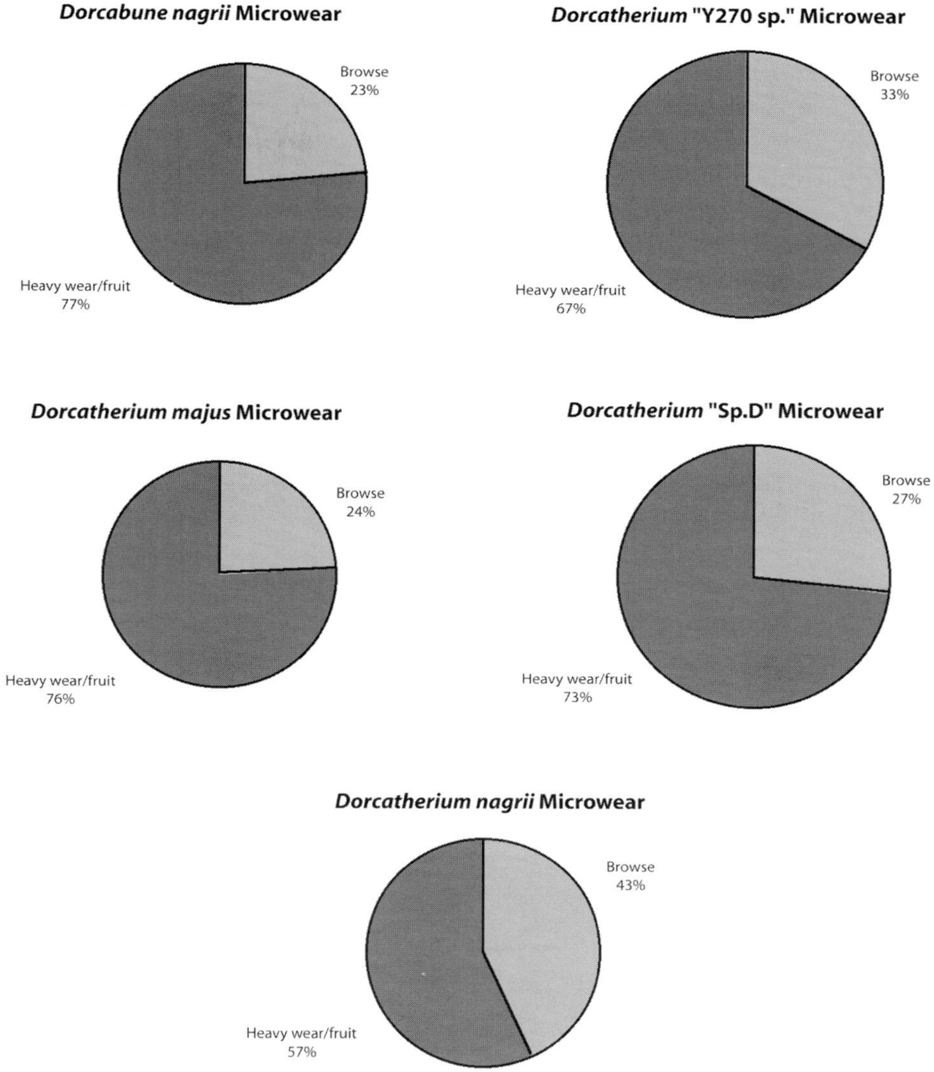

Figure 2.4 Microwear patterns for Siwalik tragulids, like modern tragulids, span a range of categories due to feeding on hard objects, probably fruit with no C_4 component. *Dorcatherium nagrii* microwear suggests the least feeding on fruits and a greater emphasis on browse.

there is a group of indeterminate small bovids (also less than 15 kg). The *Tragoceridus* specimens span 9.3–6.3 Ma and represent multiple species over time. Overall posterior probabilities for this genus fall into all dietary categories except C_3 graze due to a high number of pits and coarse scratches (see Figure 2.6, Table 2.13, and Appendix 1). Isotopic analyses (presented in Chapter Three) of *Tragoceridus* specimens 8.0 Ma or older indicate that there was not a C_4 component of the diet. Heavy wear features are therefore most likely due to feeding on hard fruits and seeds, at least prior to 8.0 Ma. More recent specimens likely incorporated C_4 graze, as indicated by a C_4 isotopic signal for a 6.3 Ma tooth (Y18210). The two specimens dated to 6.4 and 6.3 Ma show a marginally significant increase in C_4 posterior probability, from an average probability of 0.09 in older specimens to 0.29 in the two youngest ($Z = -1.94$; $p = 0.053$). These specimens also have a high fruit probability, which may suggest that fruit is still a component of the diet, or it may reflect feeding on hard-objects in general.

Table 2.7

Posterior probabilities for *Hippopotamodon sivalense*

Specimen ID	*Babyrousa*	*Hylochoerus*	*Potamochoerus*	Age (Ma)	Locality
Y18276	0.02	0.59	0.39	10.4	Y454
Y18280	0.11	0.36	0.53	10.4	Y454
Y18264	0.32	0.03	0.66	10.4	Y454
Y18261	0.55	0.01	0.44	10.4	Y454
D36	0.05	0.76	0.19	10.4	D27
Y53009	0.10	0.44	0.46	10.2	Y450
Y13972	0.13	0.03	0.83	10	Y251
Y46568	0.30	0.18	0.52	10	Y251
Y16049	0.45	0.02	0.53	10	Y251
Y41401	0.21	0.04	0.75	10	Y311
Y12861	0.22	0.32	0.46	9.3	Y309
Y4984	0.63	0.01	0.37	9.3	Y159
Y14112	0.08	0.55	0.37	9.2	Y260
Y27915	0.26	0.13	0.62	9.2	Y260
Y13524	0.20	0.35	0.45	9.2	Y239
Y5276	0.52	0.11	0.37	9.2	Y211
Y16085	0.30	0.20	0.51	9.2	Y182
Y17937	0.05	0.59	0.36	8.5	KL12
YY4086	0.11	0.56	0.34	8.3	Y174
Y330	0.05	0.39	0.56	8	Y017
Y50457	0.06	0.61	0.33	7.8	Y947
Y50019	0.34	0.04	0.62	7.8	Y917
Y50134	0.42	0.02	0.56	7.2	Y932
AVERAGE	**0.24**	**0.27**	**0.49**		
STDEV	**0.18**	**0.25**	**0.15**		
Prior to 8.4 Ma Avg	**0.25**	**0.26**	**0.49**		
After 8.4 Ma Avg	**0.19**	**0.32**	**0.48**		

Posterior probabilities for *Hippopotamodon sivalense* indicate a diet of predominantly fruit and roots, with fruit available year-round. No difference in diet is found between those specimens present before *Sivapithecus*'s extinction and those present afterwards. $N = 23$.

Table 2.8

Posterior probabilities for *Dorcabune nagrii*

Specimen ID	Browser	C_4 graze	Mixed	Fruit	C_3 graze	Age (Ma)	Locality
Y16071	0.20	0.31	0.29	0.19	0.01	10.4	Y259
Y14867	0.07	0.05	0.23	0.62	0.03	10.3	Y395
Y14351	0.46	0.06	0.16	0.30	0.02	10.2	Y450
Y11499	0.45	0.09	0.25	0.21	0.01	10.0	Y311
Y30743	0.03	0.15	0.70	0.11	0.01	10.0	Y311
Y15568	0.37	0.09	0.32	0.21	0.01	9.6	Y462
Y6857	0.05	0.24	0.39	0.29	0.03	9.3	Y227
Y10220	0.03	0.08	0.20	0.63	0.05	9.3	Y309
Y726	0.47	0.06	0.29	0.17	0.01	8.1	Y36
Y51962	0.16	0.07	0.64	0.13	0.01		Y1005
AVERAGE	**0.23**	**0.12**	**0.35**	**0.29**	**0.02**		
STDEV	**0.19**	**0.09**	**0.18**	**0.19**	**0.02**		

Posterior probabilities for *Dorcabune nagrii* indicate a diet of predominantly fruit. Isotopic evidence indicate a C_4 component is unlikely, and heavy wear in the C_4 graze and mixed diet categories probably reflects feeding on hard fruit and seeds. $N = 10$.

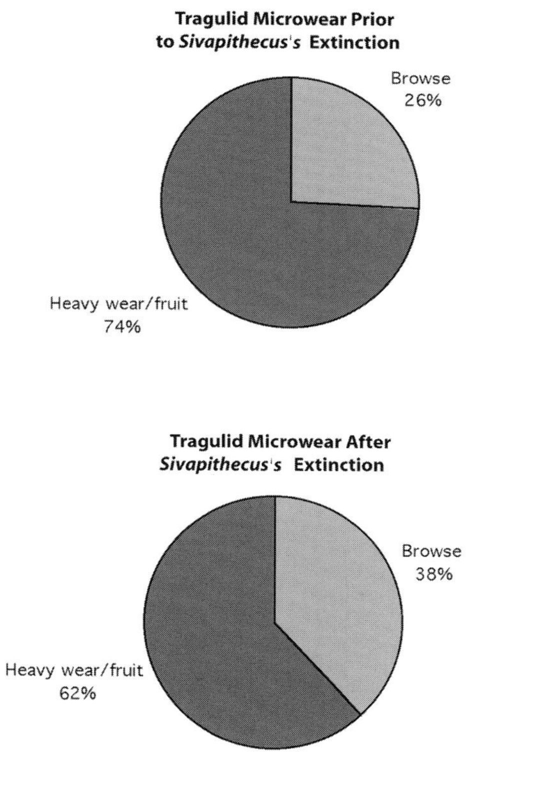

Figure 2.5 Microwear suggests that tragulids diets in general incorporated more browse after *Sivapithecus*'s extinction. $N = 65$ prior to the extinction of *Sivapithecus*. $N = 16$ afterwards.

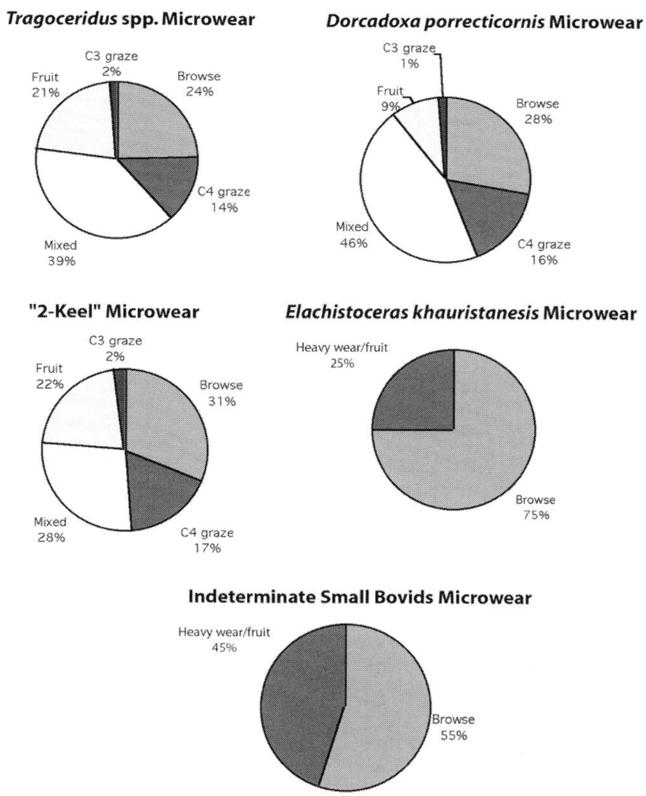

Figure 2.6 Some *Tragoceridus* species probably fed on fruit. *Dorcadoxa* and "2-Keel" fed on browse and C_4 graze. Small bovids, including *Elachistoceras* fed on fruit and high-quality browse, indicative of moist forest.

Table 2.9

Posterior probabilities for *Dorcatherium majus*

Specimen ID	Browser	C_4 graze	Mixed	Fruit	C_3 graze	Age (Ma)	Locality
D482	0.39	0.08	0.32	0.21	0.01	10.4	D027
Y15126	0.36	0.11	0.50	0.01	0.01	10.1	Y450
Y16847	0.35	0.15	0.28	0.21	0.01	10.1	Y450
Y18007	0.07	0.22	0.57	0.13	0.01	10.1	Y450
Y50639	0.19	0.19	0.58	0.03	0.01	10.1	Y450
Y50647	0.07	0.08	0.69	0.12	0.04	10.1	Y450
Y11368	0.08	0.18	0.58	0.14	0.01	10.0	Y311
Y11391	0.03	0.17	0.70	0.09	0.01	10.0	Y311
Y17707	0.12	0.32	0.45	0.09	0.01	10.0	Y311
Y17748	0.32	0.17	0.38	0.13	0.01	10.0	Y311
Y19949	0.70	0.03	0.23	0.02	0.02	10.0	Y311
Y20169	0.16	0.15	0.51	0.09	0.09	10.0	Y311
Y20552	0.34	0.12	0.31	0.22	0.01	10.0	Y311
Y22082	0.36	0.11	0.19	0.33	0.02	10.0	Y311
Y22083	0.12	0.12	0.67	0.04	0.04	10.0	Y311
Y27827	0.12	0.07	0.65	0.14	0.01	10.0	Y311
Y41417	0.23	0.03	0.65	0.06	0.03	10.0	Y311
Y41418	0.50	0.04	0.43	0.01	0.01	10.0	Y311
Y47503	0.08	0.12	0.68	0.12	0.01	10.0	Y311
Y47533	0.05	0.19	0.64	0.11	0.02	10.0	Y311
Y47604	0.04	0.35	0.49	0.11	0.02	10.0	Y311
Y47671	0.13	0.09	0.63	0.14	0.01	10.0	Y311
Y47680	0.44	0.13	0.33	0.09	0.01	10.0	Y311
Y5567	0.07	0.05	0.75	0.12	0.01	9.3	Y310
Y49530	0.47	0.07	0.28	0.17	0.01	8.0	Y888
Y49555	0.21	0.08	0.30	0.39	0.01	8.0	Y539
Y16236	0.04	0.13	0.68	0.13	0.02	8.0	Y540
Y50441	0.32	0.08	0.24	0.34	0.02	7.8	Y947
Y52556	0.25	0.08	0.28	0.37	0.02	7.3	Y941
Y19707	0.19	0.02	0.65	0.14	0.01		
Y52833	0.19	0.16	0.56	0.04	0.05		Y981
Y52856	0.70	0.01	0.26	0.03	0.00		L084
AVERAGE	**0.24**	**0.12**	**0.48**	**0.14**	**0.02**		
STDEV	**0.18**	**0.08**	**0.17**	**0.10**	**0.02**		
Prior to 8.4 Ma Avg	**0.22**	**0.14**	**0.51**	**0.12**	**0.02**		
After 8.4 Ma Avg	**0.26**	**0.09**	**0.36**	**0.28**	**0.01**		

Posterior probabilities for *Dorcatherium majus* suggest a diet of predominantly fruit. Isotopic evidence indicates no C_4 component, and the probabilities in the C_4 graze and mixed categories probably reflect feeding on fruit instead. $N = 32$.

Table 2.10

Posterior probabilities for *Dorcatherium* "Sp. D"

Specimen ID	Browser	C_4 graze	Mixed	Fruit	C_3 graze	Age (Ma)	Locality
Y14350	0.29	0.10	0.24	0.36	0.01	10.2	Y450
Y16837	0.32	0.06	0.23	0.37	0.02	10.2	Y450
Y18005	0.12	0.06	0.69	0.13	0.01	10.2	Y450
Y18006	0.33	0.08	0.39	0.20	0.01	10.2	Y450
Y6358	0.20	0.14	0.27	0.37	0.02	10.0	Y311
Y12050	0.22	0.09	0.48	0.20	0.01	10.0	Y311
Y17580	0.18	0.05	0.63	0.13	0.01	10.0	Y311
Y17705	0.22	0.24	0.30	0.23	0.01	10.0	Y311
Y20127	0.54	0.03	0.27	0.16	0.00	10.0	Y311
Y22087	0.04	0.06	0.21	0.65	0.04	10.0	Y311
Y22158	0.12	0.16	0.57	0.13	0.01	10.0	Y311
Y26227	0.46	0.04	0.47	0.02	0.01	10.0	Y311
Y41425	0.07	0.08	0.72	0.13	0.01	10.0	Y311
Y46284	0.21	0.03	0.68	0.05	0.03	10.0	Y311
Y47591	0.20	0.18	0.47	0.14	0.01	10.0	Y311
Y47658	0.08	0.11	0.68	0.12	0.01	10.0	Y311
Y47686	0.77	0.27	0.17	0.01	0.02	10.0	Y311
Y13671	0.62	0.04	0.29	0.02	0.05	9.3	Y350
Y10777	0.23	0.13	0.24	0.38	0.01	9.3	Y310
Y10780	0.33	0.04	0.42	0.21	0.00	9.3	Y310
Y11110	0.36	0.03	0.38	0.22	0.00	9.2	Y314
Y4603	0.07	0.06	0.74	0.12	0.01	8.9	Y193
Y47116	0.10	0.06	0.70	0.14	0.01	8.4	Y535
Y47117	0.08	0.23	0.54	0.14	0.01	8.4	Y535
Y5049	0.10	0.15	0.61	0.13	0.02	8.3	Y028
Y236	0.74	0.02	0.20	0.02	0.03	8.0	Y017
Y50210	0.55	0.07	0.22	0.16	0.00	7.3	Y935
Y50997	0.35	0.11	0.19	0.33	0.02	7.3	Y935
Y50999	0.12	0.20	0.33	0.32	0.04	7.3	Y935
Y52868	0.12	0.13	0.61	0.13	0.01	7.3	Y935
Y50403	0.42	0.02	0.49	0.05	0.03	7.3	Y941
AVERAGE	**0.28**	**0.10**	**0.43**	**0.19**	**0.02**		
STDEV	**0.20**	**0.07**	**0.19**	**0.14**	**0.01**		
Prior to 8.4 Ma Avg	**0.26**	**0.10**	**0.45**	**0.19**	**0.01**		
After 8.4 Ma Avg	**0.34**	**0.10**	**0.38**	**0.16**	**0.02**		

Posterior probabilities for *Dorcatherium* "sp. D" indicate a diet of predominantly fruit. Isotopic evidence suggests the C_4 graze and mixed diet categories probably reflect wear due to hard fruits and seeds rather than C_4 phytoliths. $N = 31$.

Table 2.11

Posterior probabilities for *Dorcatherium* "Y270"

Specimen ID	Browser	C_4 graze	Mixed	Fruit	C_3 graze	Age (Ma)	Locality
Y15576	0.326	0.078	0.212	0.361	0.023	9.6	Y426
Y7458	0.343	0.131	0.173	0.322	0.031	9.3	Y245
AVERAGE	**0.326**	**0.105**	**0.193**	**0.342**	**0.027**		
STDEV	**0.012**	**0.037**	**0.028**	**0.028**	**0.006**		

Posterior probabilities for *Dorcatherium* "Y270" indicate a diet of fruit with browse contributing about 30%. Isotopic evidence indicates that the C_4 and mixed diet components most likely reflect feeding on hard fruits, not graze. *N* = 2.

Table 2.12

Posterior probabilities for *Dorcatherium nagrii*

Specimen ID	Browser	C_4 graze	Mixed	Fruit	C_3 graze	Age (Ma)	Locality
Y17619	0.35	0.09	0.20	0.34	0.02	9.2	Y260
Y13771	0.72	0.03	0.20	0.01	0.04	8.7	Y166
Y13772	0.70	0.04	0.23	0.03	0.00	8.7	Y166
Y17950	0.04	0.12	0.71	0.11	0.01	8.5	Y604
Y16204	0.43	0.02	0.49	0.04	0.02	8.4	Y535
Y16205	0.28	0.05	0.57	0.06	0.04	8.4	Y535
Y16207	0.42	0.06	0.18	0.31	0.02	8.4	Y535
Y18396	0.58	0.05	0.34	0.02	0.01	8.1	Y605
Y49544	0.73	0.01	0.23	0.03	0.00	7.9	Y399
Y16402	0.53	0.05	0.40	0.02	0.01	7.9	KL11
Y51936	0.02	0.14	0.16	0.64	0.05		Y992
AVERAGE	**0.44**	**0.06**	**0.34**	**0.15**	**0.02**		
STDEV	**0.25**	**0.04**	**0.18**	**0.20**	**0.02**		
Prior to 8.4 Ma Avg	**0.42**	**0.06**	**0.37**	**0.13**	**0.02**		
After 8.4 Ma Avg	**0.61**	**0.04**	**0.32**	**0.02**	**0.01**		

Posterior probabilities for *Dorcatherium nagrii* suggest a diet highest in browse relative to other Siwalik tragulids. Furthermore, browse increases after the extinction of *Sivapithecus*. Isotopic evidence indicates that any feeding on C_4 graze was unlikely, and probabilities in the C_4 and mixed categories probably represent feeding on hard fruits and seeds. *N* = 11.

Table 2.13

Posterior probabilities for *Tragoceridus* spp.

Specimen ID	Browser	C_4 graze	Mixed	Fruit	C_3 graze	Age (Ma)	Locality
Y11593	0.21	0.15	0.51	0.13	0.01	9.3	Y312
Y5127	0.17	0.05	0.42	0.33	0.03	9.2	Y211
Y15012	0.65	0.02	0.29	0.03	0.01	8.9	Y476
Y14882	0.10	0.14	0.61	0.14	0.01	8.6	Y389
Y2349	0.30	0.11	0.45	0.13	0.01	8.1	Y024
Y18149	0.15	0.31	0.31	0.23	0.01	6.4	D013
Y18210	0.02	0.27	0.14	0.53	0.04	6.3	Y453
Y9196	0.34	0.09	0.41	0.15	0.01		Y256
AVERAGE	**0.24**	**0.14**	**0.39**	**0.21**	**0.02**		
STDEV	**0.20**	**0.10**	**0.15**	**0.16**	**0.01**		
Prior to 8.4 Ma Avg	**0.28**	**0.09**	**0.46**	**0.16**	**0.02**		
After 8.4 Ma Avg	**0.16**	**0.23**	**0.3**	**0.29**	**0.02**		

Posterior probabilities for *Tragoceridus* spp. indicate a diet that contained some fruit. The youngest specimens indicate some C_4 graze in their diets. N=8.

Thus older *Tragoceridus* species were likely to have included a fruit component in their diet, while the youngest species may have incorporated some C_4 graze.

Dorcadoxa microwear is characterized by both fine and coarse scratches, but predominantly fine (see Appendix 1). Posterior probabilities indicate a predominantly mixed browse/C_4 graze diet (see Table 2.14). The two specimens dated to 8.0 Ma or older had more browse in their diets than the three specimens dated to 7.3 Ma or younger ($Z = -1.73$; $p = 0.083$), while the C_4 component increased in the younger three ($Z = -1.73$; $p = 0.083$) (see Figure 2.7). Isotopic analyses of *Dorcadoxa* specimens older than 8.0 Ma indicate a pure C_3 diet, thus also suggesting C_4 became a stronger component after 8.0 Ma. Fruit probability also showed a marginally significant increase in the younger three ($Z = -1.73$; $p = 0.083$), but given the overall evidence for a mixed browse/C_4 graze diet, this fruit probability is probably a reflection of increase in wear due to phytoliths, combined with a very small sample size.

Microwear was only available for two "2-Keel" teeth, both dating to 7.3 Ma. These teeth suggest, like *Dorcadoxa*, a mixed browse/C_4 graze diet (see Table 2.15). Again, the fruit probabilities suggest a fruit component, which alternatively could be a reflection of phytoliths and a very small sample size. Given the low number of pits on these specimens and the large body size of this taxon, any fruit component was probably small.

Elachistoceras was also represented by only two specimens, both dated prior to the extinction of *Sivapithecus*. These two specimens have only fine scratches, with a low scratch count, suggesting a predominantly browse diet (0.75), although some evidence of hard-object eating is suggested by the posterior probability of a mixed diet (0.20) (see Table 2.16 and Figure 2.6). Given its small body size, *Elachistoceras* must have been feeding on young, protein-rich shoots and leaves, like many small forest antelope and duikers today (Kingdon 1997).

Among the indeterminate small bovids, most specimens are characterized by a combination of fine scratches and a high number of pits, including large pits. These features are similar to the combination seen in chimpanzees, suggesting a diet of soft fruit. Posterior probabilities are highest in the browse (0.55) and mixed diet (0.32) categories (see Table 2.17). Given their very small

Table 2.14

Posterior probabilities for *Dorcadoxa porrecticornis*

Specimen ID	Browser	C_4 graze	Mixed	Fruit	C_3 graze	Age (Ma)	Locality
Y12367	0.40	0.04	0.50	0.04	0.02	8.9	Y325
Y2817	0.81	0.01	0.16	0.02	0.00	8	Y017
Y50771	0.07	0.17	0.60	0.16	0.01	7.3	Y927
Y50824	0.04	0.24	0.58	0.13	0.01	7.2	Y452
Y50737	0.10	0.32	0.44	0.12	0.02	6.4	D013
AVERAGE	**0.28**	**0.16**	**0.46**	**0.09**	**0.01**		
STDEV	**0.33**	**0.13**	**0.18**	**0.06**	**0.01**		
Prior to 8.0 Ma Avg	**0.61**	**0.03**	**0.33**	**0.03**	**0.01**		
After 8.0 Ma Avg	**0.07**	**0.25**	**0.54**	**0.13**	**0.02**		

Posterior probabilities for *Dorcadoxa porrecticornis* indicate a mixed browse /C_4 graze diet, with the amount of C_4 graze increasing over time. $N = 5$.

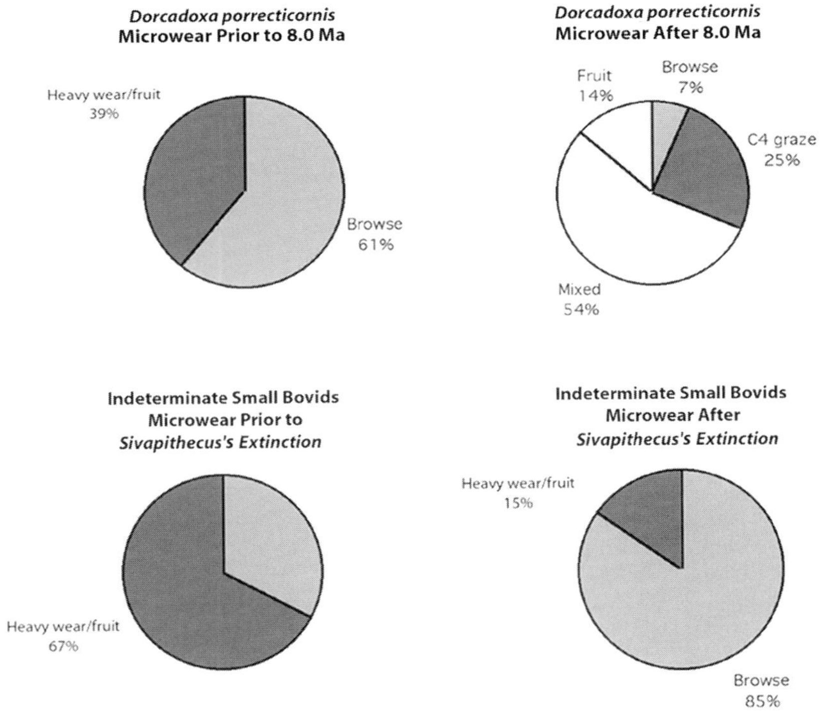

Figure 2.7 Microwear for *Dorcadoxa porrecticornis* and the indeterminate small bovids indicate shifts in diet over time, with browse being replaced by C_4 graze in *Dorcadoxa* and fruit being replaced by browse in the small bovids.

body size and isotopic values of fauna in general before 8.0 Ma, it is very unlikely these bovids fed on any C_4 graze. However, the low probabilities in the fruit category combined with the fine scratches suggests that these small Siwalik bovids were feeding on softer fruits than many of the duikers in the modern dataset. Finally, if individuals are divided into two groups with respect to *Sivapithecus*'s extinction, the three specimens dated post-*Sivapithecus* show significantly smaller probabilities for all of the hard-food categories (C_4, mixed, and fruit) (for each of the three, $Z = -2.12$; $p = 0.034$), compared to the four specimens older than 8.4 Ma (see Figure 2.7). Furthermore, the younger specimens show a marginally significant increase in browse from 0.33 probability prior to 8.4 Ma to 0.84 after 8.4 Ma ($Z = -1.77$; $p = 0.077$). These results suggest that small bovids in general may have fed substantially on fruits until some time after the extinction of *Sivapithecus*, at which point browse became a larger portion of the diet.

Finally, a comparison of the small tragulids and bovids (under 15 kg) to the larger tragulids and bovids indicates some differences in their microwear features that could represent differences in their diets or differences in mastication. The teeth of smaller ruminants, on average, have fewer cross scratches (scratches perpendicular to one another) than do larger ruminants, suggesting more unidirectional mastication ($Z = -2.57$; $p = 0.01$) (see Table 2.18). They also show a marginally significant reduction in number of scratches ($Z = -1.71$; $p = 0.088$). These differences in microwear features and possibly mastication suggest that smaller ruminants might have fed on softer fruits, requiring less grinding and resulting in less wear, than did larger ruminants.

Conclusions

The new light microscope microwear technique (Solounias and Semprebon 2002) distinguished well between known diets, allowing for diet determination for

Table 2.15

Posterior probabilities for "2-Keel"

Specimen ID	Browser	C_4 graze	Mixed	Fruit	C_3 graze	Age (Ma)	Locality
Y50399	0.22	0.20	0.25	0.31	0.03	7.3	Y941
Y50356	0.41	0.14	0.31	0.13	0.01	7.3	Y941
AVERAGE	**0.31**	**0.17**	**0.28**	**0.22**	**0.02**		
STDEV	**0.14**	**0.14**	**0.04**	**0.13**	**0.01**		

Posterior probabilities for "2-Keel" indicate a mixed browse/C_4 graze diet. $N = 2$.

Table 2.16

Posterior probabilities for *Elachistoceras khauristanensis*

Specimen ID	Browser	C_4 graze	Mixed	Fruit	C_3 graze	Age (Ma)	Locality
Y18346	0.67	0.04	0.25	0.05	0.00	8.5	KL12
Y17090	0.83	0.01	0.15	0.01	0.00	9.5	Y262
AVERAGE	**0.75**	**0.02**	**0.20**	**0.03**	**0.00**		
STDEV	**0.11**	**0.02**	**0.07**	**0.02**	**0.00**		

Posterior probabilities for *Elachistoceras khauristanensis* indicate a diet of browse with a fruit component. Wear in the mixed category can be interpreted as fruit due to isotopic faunal analyses. $N = 2$.

fossil taxa, including changes in food choice over time. Microwear results indicate that *Sivapithecus* was similar to modern large-bodied frugivorous apes, with some hard fruits or seeds incorporated into its diet, but it is not closely similar to any single modern species. Other fossil taxa contemporaneous with *Sivapithecus* were also frugivorous, including two suids with diets similar to the African red river hog today, tragulids similar to modern counterparts, some small bovids which probably fed on soft fruits in combination with young leaves and shoots, and one larger bovid, *Tragoceridus*, in which older species were predominantly frugivorous. Determination of the fruit component in these fossil taxa was somewhat ambiguous, for, as in modern frugivores, hard fruits and seeds yield microwear similar to that caused by C_4 phytoliths. However, when microwear is combined with isotopic evidence from mammalian teeth and paleosols, the C_4 /mixed diet signals for many of the older specimens in this study can be interpreted as evidence of frugivory.

There is no evidence for C_3 graze among any of the taxa analyzed in this study, but fauna likely to have fed in open habitat, such as equids, anthracotheres, and rhinoceroses, were not included.

Propotamochoerus microwear results suggest that its preferred fruits varied in availability geographically or seasonally. Localities represented by multiple specimens with different diets support seasonal variability. Several tragulid specimens with high probabilities in the browse category also suggest periods of fruit shortages. Furthermore, a decrease in fruit probabilities among some tragulids and bovids suggest that fruit availability decreased around the time *Sivapithecus* went extinct. Bovid microwear indicates that some of the forest once inhabited by frugivores became replaced with open habitat, including patches of C_4 grass. While fruit availability may have been seasonal, the presence of bovids under 15 kg suggests that non-fibrous, high-protein browse was available year-round, indicative of moist forest with continuous plant production well after *Sivapithecus* became extinct. The frugivorous suid species provide further evidence for moist forest, for the modern analogues *Potamochoerus* and *Babyrousa* inhabit moist forest, with *Potamochoerus* rarely found outside rainforest (Kingdon 1997; Nowak 1999).

The large proportion of Siwalik fauna that were frugivorous suggests that *Sivapithecus* habitat preference was similar to that of modern apes in fruit abundance,

Table 2.17

Posterior probabilities for small indeterminate bovids

Specimen ID	Browser	C_4 graze	Mixed	Fruit	C_3 graze	Age (Ma)	Locality
Y16836	0.60	0.06	0.29	0.04	0.01	10.2	Y450
Y46345	0.40	0.16	0.38	0.06	0.01	9.9	Y870
Y46348	0.09	0.20	0.59	0.11	0.02	9.9	Y870
Y14860	0.23	0.04	0.65	0.05	0.04	8.7	Y388
Y542	0.76	0.04	0.16	0.01	0.02	8.1	YO24
Y16978	0.85	0.01	0.11	0.01	0.02	8	Y545
Y14089	0.91	0.01	0.07	0.01	0.01	7.1	Y374
AVERAGE	**0.55**	**0.07**	**0.32**	**0.04**	**0.02**		
STDEV	**0.32**	**0.08**	**0.23**	**0.04**	**0.01**		
Prior to 8.4 Ma Avg	**0.33**	**0.11**	**0.48**	**0.07**	**0.02**		
After 8.4 Ma Avg	**0.84**	**0.02**	**0.11**	**0.01**	**0.02**		

Posterior probabilities for a group of indeterminate bovids. Specimens indicate a shift from fruit to browse over time. Mixed and C_4 categories can be interpreted as fruit due to isotopic faunal analyses. $N = 7$.

although it is unknown whether it was similar in terms of seasonal fruit availability. Loss of *Sivapithecus* and two frugivorous tragulids, combined with shifts from fruit to browse in some small bovids, and shifts from browse to graze in larger bovids, suggests that forests contributed a smaller proportion to the vegetation mosaic over time as open habitat, including patches of C_4 grass became more abundant. Continual presence of frugivorous suids and very small bovids, however, suggests that forests were still present well after the extinction of *Sivapithecus*. If loss of forest resulted in the extinction of *Sivapithecus* and some tragulids, it was likely a fragmentation of forest rather than a complete loss of forest.

Table 2.18

Microwear differences between small versus large bovids and tragulids

Taxon	Cross Scratches (%)	Large Pits (%)	Gouges (%)	Number of Pits	Number Scratches
Dorcatherium nagrii	27.3	81.8	18.2	26.8	11.3
Dorcatherium Y270	0	100	0	25.5	11
Elachistoceras	0	0	0	22.8	7.3
Indet. Small bovids	14.3	57.1	0	24.8	12.5
AVERAGE	**10.40**	**59.73**	**4.55**	**24.98**	**10.53**
Dorcabune nagrii	80	100	50	30.4	15.8
D. majus	68.8	81.3	28.1	27.8	17.3
Dorcatherium Sp. D	61.3	93.5	29	34	15.4
Tragoceridus	50	50	12.5	26.3	13.6
Dorcadoxa	60	40	0	22.6	16.6
"2-Keel"	50	50	0	14.8	6.8
AVERAGE	**61.68**	**69.13**	**19.93**	**25.98**	**14.25**

Small tragulids and bovids (*Dorcatherium nagrii, Dorcatherium* "Y270," *Elachistoceras*, and indeterminate small bovids) differ from larger tragulids and bovids (*Dorcabune nagrii, D. majus, Dorcatherium* "Sp. D.," *Tragoceridus* spp., *Dorcadoxa*, and "2-Keel") in having fewer cross scratches and fewer scratches in general. These differences suggest more unidirectional mastication on less abrasive foods, such as softer fruits.

3 Reconstruction for Two Time Intervals—9.3–9.2 Ma and 8.1–8.0 Ma—from Isotopic Analyses of Tooth Enamel

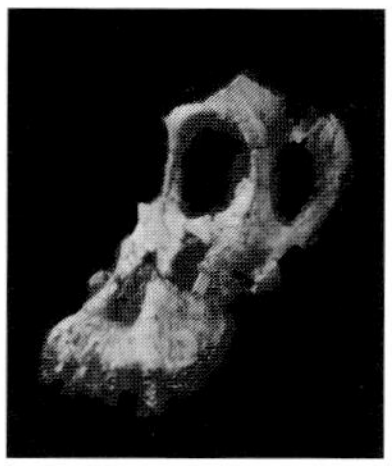

Introduction

Stable isotopic analyses of tooth enamel provide a useful means for reconstructing diets, habitats, and paleoclimates. This study uses carbon and oxygen stable isotopic compositions of inorganic carbonate in enamel from various taxa to reconstruct faunal paleodiets and habitats for two time intervals—a period when *Sivapithecus* material is abundant (9.3–9.2 Ma) and an interval shortly after *Sivapithecus* went extinct (8.1–8.0 Ma). The purpose of this study is to determine the composition of the vegetation mosaic while *Sivapithecus* was present, and what changes in habitat and climate may have taken place by the time *Sivapithecus* went extinct.

Analyses of modern herbivores have shown that the carbon isotopic composition of body tissues reflects the carbon composition of vegetation eaten during tissue formation (DeNiro and Epstein 1978; Van der Merwe and Vogel 1978; Van der Merwe 1982). Likewise, tissue oxygen isotope ratios provide information on the oxygen composition of water intake during development (Longinelli 1984; Luz et al. 1984; Luz and Kolodny 1985). Tooth enamel in particular has been used to reconstruct paleodiets, habitats, and climatic change (Lee-Thorp et al. 1989; Cerling et al. 1997a), for unlike fossilized bone, it is far less susceptible to diagenesis (Ayliffe et al. 1994; Koch et al. 1997; Quade et al. 1992; Wang and Cerling 1994).

Carbon isotopes are informative for both paleodietary and paleohabitat reconstructions due to differences in carbon assimilation between different plants. Most terrestrial plants assimilate atmospheric CO_2 by one of two photosynthetic pathways. C_3 plants, which include almost all trees and shrubs, and only those grasses favored by cool, wet growing seasons, utilize the enzyme ribulose biphosphate carboxylase-oxygenase (Rubisco) to fix CO_2, forming a three-carbon sugar. C_4 plants, which include exclusively grasses and sedges growing in hot, dry habitats, use a different enzyme to fix CO_2, phosphoenolpyruvate (PEP) carboxylase, resulting in a four-carbon acid (Farquhar et al. 1989). These two photosynthetic pathways yield greatly different C values ($\delta^{13}C$ is the ratio of ^{13}C to ^{12}C relative to a standard, in this case Pee Dee Belemnite or PDB), with C_3 plants falling between –22 and –35‰ (per mil), and C_4 plants ranging between –10 and –15‰ (Bender 1971; Vogel 1980). A third photosynthetic pathway, known as CAM plants (crassulean acid metabolism) is ignored in this study, for these plants are xeric-adapted desert succulents not likely to have been present in significant abundance in the Siwaliks.

The non-overlapping $\delta^{13}C$ values of C_3 versus C_4 plants allows discrimination of browsing versus grazing diet from analyses of tooth enamel beginning with the first appearance of C_4 grasses in the Late Miocene. Enamel bioapatite records the carbon signal obtained from vegetation eaten, with an enrichment in $\delta^{13}C$ due to fractionation from metabolic processes. The enrichment factor for all taxa analyzed in this Siwalik study is assumed to be 14‰, as found by Harris and Cerling (1999) in an extensive analysis of ungulates. It should be noted, however, that these ungulates were all ruminant foregut fermenters, and suids may have slightly lower enrichment factors due to a different digestive system, though this remains to be determined. Thus, enamel with a $\delta^{13}C$ of –14‰ for any Siwalik taxa is interpreted as indicating plant diet averaging –28‰, or C_3 browse.

The C_3/C_4 distinction is useful for paleoenvironmental as well as dietary inferences, for the C_4 photosynthetic pathway is energetically expensive compared to the C_3 pathway and is only competitively advantageous in hot, dry climates. Carbon dioxide enters a leaf through its stomata, but stomata must remain closed much of the time to conserve water in hot, dry climates. The enzyme used by C_4 plants has a higher affinity for CO_2 than does the C_3 enzyme, thus allowing the plant to take up more CO_2 when the stomata are open (Ehleringer et al. 1991). Furthermore, this concentrated version yields a ratio of CO_2 to O_2 high enough that little CO_2 is lost to photorespiration. These energetically expensive adaptations allow C_4 plants to function better than C_3 plants under high moisture stress during warm/hot growing seasons. However, because of the energetic tradeoff, C_4 plants are restricted to warm, relatively dry regions, with C_3 plants competitively superior in all other habitats.

Not only is the distinction between C_3 and C_4 significant, but the wide range of $\delta^{13}C$ values displayed by C_3 plants is an indication of vegetation type as well. Variability in $\delta^{13}C$ values among C_3 plants is a response to variation in irradiance and moisture stress as well as to CO_2 trapping under forest canopies (Ehleringer et al. 1986, 1987; Quade et al. 1995b; Van der Merwe and Medina 1989). Ehleringer et al. (1986, 1987) studied monsoon forests in China and found that plants in a relatively closed canopy versus an open canopy differed in $\delta^{13}C$ values by several per mil on average. For any given plant species, individuals growing in open habitats were consistently more enriched in carbon values than those growing under closed canopies, with those growing under intermediate coverage having intermediate $\delta^{13}C$ values. Habitat alone does not explain all of the variance, however. Ehleringer (1986) found that among ten species, a range of 2–3‰ in carbon values was found for any given habitat (closed or open). When over 100 plant species were sampled in a subsequent study (Ehleringer 1987 et al.), a range of 6–7‰ was found within a given habitat, though again, each species showed consistent enrichment with decreasing coverage. Thus $\delta^{13}C$ values of plants respond to differences in habitat, but habitat alone does not explain the entire variability of carbon values observed for C_3 plants.

Among herbivores feeding in different habitats (closed versus open), $\delta^{13}C$ values of tooth enamel respond to the differences in vegetation. Van der Merwe and Medina (1989) demonstrated that in a Brazilian rainforest, fauna foraging in forest openings showed a 2‰ enrichment in bone collagen relative to fauna foraging in relatively closed forest. Furthermore, in dense closed canopy forests in which little light penetrates to the ground, and exchange of CO_2 in forest floor air with open tropospheric air is inhibited, carbon values for both the vegetation and fauna feeding there are even more depleted (Cerling et al 1997a). Medina et al. (1986) reported vegetation $\delta^{13}C$ values as low as −36‰ for forest floor vegetation in Venezuela, compared to values as high as −29‰ at the upper canopy, while Cerling and Harris (2002) have found forest hog (*Hylochoerus*) enamel from the Ituri forest as depleted as −23‰, meaning it ate vegetation as depleted as −37‰. Thus $\delta^{13}C$ variability in C_3 plants translates into variability among fauna. Furthermore, very closed canopy forests yield a uniquely depleted $\delta^{13}C$ signal that should be detectable in the fossil record.

This range in $\delta^{13}C$ values for C_3 browse has in fact been used in reconstructing paleoenvironments. Quade et al. (1995b) have found consistent differences in $\delta^{13}C$ values between different Miocene browsers at Pasalar, a 15 Ma site in Turkey, suggesting that some taxa fed predominantly in open habitats, while others fed in closed habitats. For example, *Giraffokeryx* (a giraffid) and *Capotragoides* (a bovid) yielded the most depleted $\delta^{13}C$ values among fossil mammals from Pasalar (consistently lighter than −12‰ [PDB]), suggesting they fed in well forested areas with vegetation sources averaging −26‰. On the other hand, *Conohyus* (a suid) and *Hypsodontus* (a bovid) yielded enriched $\delta^{13}C$ values (around −10‰), implying that they preferred more open habitat plant foods. Likewise, an isotopic study of fauna from Fort Ternan, a 14 Ma Kenyan site, yielded $\delta^{13}C$ values ranging from −13 to −8.6‰, suggested a woodland/forest habitat with no closed canopy (Cerling et al. 1997a). In the Siwaliks of Pakistan, carbon isotopes of paleosols, tooth enamel, and fossil avian eggshells have been used to reconstruct paleohabitats. Paleosol $\delta^{13}C$ ratios reflect the isotopic composition of vegetation growing above the site of soil formation, with about a 15‰ enrichment due to fractionation effects of gaseous diffusion and equilibrium exchange (Quade and Cerling 1995, Quade et al. 1995a). Thus, soil carbonate formed in the presence of pure C_3 vegetation would have an average $\delta^{13}C$ (PDB) value of around −13‰, while pure C_4 biomass would yield values around +2‰. $\delta^{13}C$ values above −9‰ may indicate a small amount of C_4 (Quade and Cerling 1995; Quade et al. 1989). Paleosol carbonates in the Siwaliks indicate a relatively rapid shift in carbon values during the Late Miocene (Quade and Cerling 1995; Quade et al. 1989, 1992). The results were originally calibrated to the Berggren et al. (1985) timescale but have since been recalculated by Barry et al. (2002) using the Cande and Kent (1995) timescale. The beginning of the carbon shift from C_3 to C_4 vegetation occurs around 8.1 Ma, and the biomass becomes pure C_4 by 5.9 Ma. However, this shift is complex, with depleted values even after 7.4 Ma, indicating patches of C_3, and more enriched values indicating pure C_4 grasslands as early as 6.8 Ma.

Previous tooth enamel and eggshell data have also shown this carbon transition from C_3 browse to mixed and C_4 grazing diets, with a distinct shift in $\delta^{13}C$ values after 8 Ma (adjusted to Cande and Kent 1995), similar to the paleosol data (Morgan et al. 1994; Stern et al. 1994). These studies yield $\delta^{13}C$ values as enriched as –8 to –9‰ before the 8 Ma transition, suggesting C_4 graze may have been a very minor component before 8 Ma. It is possible that if there were early C_4 grasses, they simply were not sampled in the paleosol record, or soils associated with this vegetation may not have preserved the carbonate nodules necessary for isotopic analyses. In addition to carbon isotopes, oxygen isotopes are also useful in reconstructing diets and habitats as well as paleoclimatic change, but they have received less attention from paleontologists because their interpretation is not straightforward. The $\delta^{18}O$ value of body water, and hence tooth enamel, is determined largely by the oxygen isotopic composition of ingested water, both in drinking water and in plants eaten, with a minor contributing component being metabolic processes which vary among species (Luz and Kolodny 1985; Luz et al. 1984). The $\delta^{18}O$ value of environmental water, in turn, is controlled by the $\delta^{18}O$ value of local precipitation, with water sources then potentially later modified by evaporative effects (Longinelli 1984; Quade et al. 1995b).

Due to evaporative effects, certain caveats must apply when interpreting oxygen ratios. These include differences between depleted permanent water sources and enriched ephemeral ones, differences between depleted well-shaded sources and enriched open sources, and potential similarities between depleted permanent sources and depleted well-shaded ones. Furthermore, there are differences between obligate drinkers versus animals that obtain all of their water from foliage. Herbivores that obtain most of their water from plant sources have oxygen values reflecting not only $\delta^{18}O$ values of local rainwater, but also environmental relative humidities which affect oxygen enrichment in plant leaves via evapotranspiration (Ayliffe et al. 1992). A sample across different Australian climates of macropods, known to obtain at least 40 percent of their ingested water from vegetation, yielded a range in $\delta^{18}O$ values of bone phosphate that was greater than the range of values for Australian rainwater (Ayliffe and Chivas 1990). Thus differences in $\delta^{18}O$ values among fauna can often be interpreted in several different ways, including differences in habitat and differences in water dependence.

Nonetheless, changes in $\delta^{18}O$ values over time can suggest changes in paleoclimates, and consistent differences in $\delta^{18}O$ values between taxa are useful for inferring closed versus open habitat preferences. For example, both Cerling et al. (1997a) and Quade et al. (1995b) found that Miocene giraffids at Fort Ternan and Pasalar yielded very enriched oxygen values compared to contemporaneous fauna, consistent with feeding in dry, open canopy tops. These enriched values are also consistent with diet with feeding habits of modern giraffes, which obtain much of their water from canopy foliage (Quade et al. 1995b; Zazzo 2000). Furthermore, oxygen values at Pasalar, like the carbon values, were consistently different between some taxa, suggesting some were drinking in open habitats, others, in well-shaded areas (Quade et al. 1995b). For example, *Capotragoides*, which had very depleted $\delta^{13}C$ values, also yielded the most depleted $\delta^{18}O$ values, suggesting it fed in well-shaded vegetation. *Hypsodontus* and *Conohyus* were two taxa with enriched $\delta^{13}C$ values, and *Hypsodontus* displays enriched oxygen values as well, suggesting that it fed and drank in open habitat, while the $\delta^{18}O$ values for *Conohyus* are mixed, suggesting possibly a range of habitats.

In addition to reconstructing vegetation mosaics via inter-species variability, changes in $\delta^{18}O$ values over evolutionary time suggest changes in climate, or more specifically, changes in temperature or rainfall regime. The $\delta^{18}O$ value of precipitation is controlled by a number of factors, including ambient temperature and amount of rainfall. In temperate regions, oxygen values respond to seasonal variability in temperature, for temperature varies more on an annual basis than does amount of rainfall. In temperate regions, renewal of vapor from the air by condensation is affected by cooling; therefore, oxygen isotopes can track changes in temperature due to differences in evaporative pressure between ^{18}O and ^{16}O (Yurtsever and Gat 1981). In the tropics, where rainfall is more variable than temperature, oxygen values track changes in rainfall, for ^{18}O precipitates more readily than ^{16}O; therefore, the more it rains, the less ^{18}O is present in atmospheric water, and the more depleted subsequent precipitation becomes (Dansgaard 1964).

Most research on changes in oxygen values over time in terrestrial records has involved studies of paleosols or fossil bivalve shells. Plants do not fractionate oxygen during water uptake; therefore, oxygen isotopic values of paleosols are determined only by the $\delta^{18}O$ value of meteoric water at the site, by the subsequent evaporation, and by temperature. In the Siwaliks of Pakistan, paleosols indicate a shift in $\delta^{18}O$ values beginning around 9.15 Ma, with samples becoming more enriched over time (Quade and Cerling 1995; Quade et al. 1989). This shift appears as suddenly as the carbon shift, but it precedes the carbon shift by 1.1 Ma. The oxygen shift also lasts longer than the carbon shift and appears to take place in phases, with most of the enrichment occurring by 8.0 Ma, and another pulse after 6.0 Ma. Quade and Cerling (1995) outlined a number of possible factors that could have resulted in the enrichment, including an increase in temperature by five degrees Celsius; a decrease in rainfall, with a reduction by 250 mm between 8.5–6 Ma; a change in the source of precipitation, with the rise of the Himalayas blocking northern storms, thus restricting rain sources to the southern, more enriched storms; or more likely, a combination of factors including a decrease in rainfall and increasing seasonality of rainfall with increasing strength of the Asian monsoon. These paleosol oxygen isotope results are compatible with foraminiferal records which show the initiation of strong summer Arabian upwellings around 9.4 Ma, with continued strengthening until around 8.2 Ma, suggesting the beginning or at least major intensification of the South Asian monsoon (Kroon et al. 1991, corrected to Cande and Kent 1995). Furthermore, Miocene deposits in Nepal show a 3‰ enrichment in oxygen isotopes of fossil bivalve shells around 7.5 Ma, also suggesting aridification (Dettman et al. 2001).

The inferred increase in seasonality, aridification, and summer rainfall suggested by Siwalik paleosol and bivalve oxygen isotopes would have favored C_4 plants, and hence more open habitats, over C_3 plants and closed forest. This Siwalik isotopic investigation of mammalian tooth enamel seeks to reconstruct the C_3/C_4 transition over two time intervals in addition to changes in the ratio of closed to open habitat and any corresponding climatic shifts.

Materials and Methods

A total of 101 fossil teeth from 17 taxa were selected from two time intervals, 9.3–9.2 Ma, referred to here as the *Sivapithecus*-level (S-level), and 8.1–8.0 Ma, or the post-*Sivapithecus* level (PS-level). These intervals were chosen because both are comparably abundant in fossils and have been well surveyed. Furthermore, much of the *Sivapithecus* material found comes from the S-level, suggesting this was a period of relative abundance for the ape, and *Sivapithecus* went extinct around 8.4 Ma, not long before the fossiliferous younger level sampled. Thus from the two time intervals, environmental inferences can be made for the habitat when *Sivapithecus* was at its prime, and again just after *Sivapithecus* went extinct. All teeth sampled were either premolars or molars, and all came from different individuals, except where noted for seven bovid specimens. Taxa represented include *Sivapithecus*, the suoids *Propotamochoerus hysudricus*, *Hippopotamodon sivalense*, *Tetraconodon magnus*, and *Schizochoerus gandakasensis*; anthracotheres (an extinct group possibly related to Hippopotamidae) *Microbunodon punjabiense* and *Hemimeryx* large sp.; proboscideans known as gomphotheres and deinotherids; a giraffid *Bramatherium megacephalum*; rhinocerotids; a chalicothere (a fossil perissodactyl with long forelimbs and short hindlimbs); the bovids *Tragoceridus* spp. and *Dorcadoxa porrecticornis*; the tragulids *Dorcabune nagrii* and *Dorcatherium majus*; and the equid genus *Hipparion*. The taxa included are all the large mammal herbivores present at the two levels that had enamel thick enough to sample. The large taxonomic range represents most of the community, with small mammals, carnivores, very small tragulids and bovids, and bovids of unknown taxonomy, excluded.

Samples of pure enamel were removed from teeth using a Dremel high-speed rotary tool with a tungsten-carbide or diamond-impregnated bit. In an initial isotopic analysis of some of the horse teeth, CO_2 was extracted from carbonate by reacting it with phosphoric acid and isolating it cryogenically. All other samples were run using an automated carbonate microsample device. For those samples extracted cryogenically, 30–50 mg of tooth enamel were ground, washed with three percent hydrogen peroxide for one hour to remove organic matter, rinsed thoroughly with distilled water, and washed with 1M acetic acid for one hour to remove surficial carbonates, followed by a thorough rinse with distilled water and finally, ethanol. They were then reacted with 100 percent phosphoric acid at 25 degrees Celsius for 48 hours, and the evolved CO_2 was subsequently isolated cryogenically, reacted with Ag wool to remove any SO_2, and analyzed on a Finnigan delta S stable isotope ratio mass spectrometer. For all other samples, only .5–.7 mg were required, and these samples were washed with three percent hydrogen peroxide for 15 minutes and rinsed, followed by .1M acetic acid for 15 minutes and rinsed. Samples were then reacted with 100 percent phosphoric acid at 90 degrees Celsius on the automated carbonate device interfaced with a Finnigan MAT 252 stable isotope ratio mass spectrometer. Isotopic ratios are presented in the per mil (‰) notation

$$\delta^{13}C \text{ (or } \delta^{18}O) = (R_{sample}/R_{PDB}-1) \times 1{,}000$$

where R_{sample} and R_{PDB} are the ratios $^{13}C/^{12}C$ (or $^{18}O/^{16}O$) in the sample and standard respectively, and the isotope reference standard is PDB. Oxygen values are reported assuming the acid-calcite fractionation factor for calcite. All samples were analyzed at the University of Utah.

While two different methods were used to obtain the equid values, when both methods were performed on the same tooth or teeth from the same jaw, they yielded comparable results in almost all cases (see Table 3.1). For Y50803, the $\delta^{18}O$ values from the two different methods could be very different due to the fact that for the cryogenic method, the sample was taken at the root of the tooth, and the root develops much later than the crown (Hillson 1986). The cryogenic method did yield slightly more depleted $\delta^{18}O$ values for some of the other teeth as well, but the difference between the two methods is generally minimal.

Results and Discussion

Carbon and oxygen results are first presented for the S-level to provide a reconstruction of the vegetation mosaic at a time when *Sivapithecus* was present. Results are then reported for the PS-level to reconstruct the vegetation mosaic just after *Sivapithecus* went extinct. Changes in habitat are then inferred from a comparison of the two time levels, and climatic changes are inferred from differences in oxygen isotopic values between the two levels.

Carbon values are interpreted based upon certain cutoff values for diets of C_3 versus C_4 vegetation as well as for closed versus open habitat. The $\delta^{13}C$ values for tooth apatite for pure C_3 modern diets in general range from approximately –20 to –8‰, with the extremes representing distinctive habitats, depleted very closed canopy rainforest and enriched very open habitat (Cerling et al. 1997a). A pure C_4 diet, on the other hand, yields $\delta^{13}C_{(apatite)}$ values from +1 to +4‰. Modern $\delta^{13}C$ values are depleted by 1.5‰ relative to Miocene values due to a shift in atmospheric carbon isotope ratios since the industrial revolution and burning of fossil fuels. Therefore, a value of –8‰ today, or the most enriched pure C_3 diets, is comparable to a Miocene value of –6.5‰. For the purposes of this analysis, a cutoff $\delta^{13}C$ value of –7.5‰ is used to distinguish evidence of at least some C_4 in the diet. In terms of habitat, $\delta^{13}C$ values below –12‰ represent closed habitat, while those above –10‰ suggest affinities for more open habitat (Quade et al. 1995b).

The S-level (9.3–9.2 Ma)

The $\delta^{13}C$ results for the S-level fall between –14.6 and –8.1‰, for a range of 6.5‰ and a mean of –11.3‰, indicating a predominantly, if not entirely, C_3 habitat (see Table 3.2 and Figure 3.1). The most enriched values from the S-level are greater than –10‰ and therefore suggest that some animals, in this case *Hipparion*, an

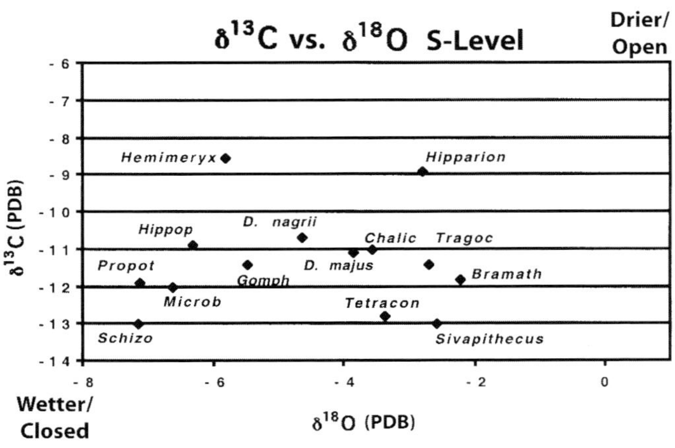

Figure 3.1. Average carbon and oxygen isotopic values for each taxon in the S-level indicate a range of feeding adaptations exploiting different parts of the habitat. Taxa include *Sivapithecus*, the suoids *Propotamochoerus*, *Schizochoerus*, *Tetraconodon*, and *Hippopotamodon*; gomphotheres; the anthracotheres *Hemimeryx* and *Microbunodon*; the giraffid *Bramatherium*; the tragulids *Dorcabune nagrii* and *Dorcatherium majus*; the bovid *Tragoceridus*; the equid *Hipparion*; and a chalicothere. *Bramatherium* is similar to modern giraffes in enriched oxygen values, indicative of feeding on upper canopy foliage. *Hemimeryx* is similar to hippopotamus today in enriched carbon but depleted oxygen. *Hipparion* also fed in open habitat. *Schizochoerus*, *Propotamochoerus*, and *Microbunodon* fed in closed, wet forest.

equid (average −8.9‰), and *Hemimeryx*, an anthracothere (−8.5‰), fed in open habitats, where plants are under high water and light stress. These results could also potentially indicate grazing on C_3 grasses if open C_3 grasslands were part of the Miocene landscape. Alternatively, the enriched values could suggest diets containing a small percentage of C_4 grass. The most depleted values, those below −12‰ (found for *Sivapithecus*, gomphotheres; the giraffid *Bramatherium*; the suoids *Propotamochoerus*, *Tetraconodon*, and *Schizochoerus*; the anthracothere *Microbunodon*; and the tragulid *D. majus*), are consistent with closed woodland or forest habitat, though none indicate a closed canopy with little light penetration or lack of exchange of forest and tropospheric CO_2.

The $\delta^{18}O$ values for the S–level fall between −9.2 and +.7‰, for a range of 9.9‰, and an average of −4.5‰. For several taxa, individuals cluster in their $\delta^{13}C$ and/or $\delta^{18}O$ values, suggesting that they fed in particular parts of the habitat. For example, *Hipparion* specimens have the most enriched carbon and oxygen values, suggesting they prefer more open, drier patches of the

Table 3.1

Comparison of cryogenic and automated carbonate microsampler methods

Specimen	Samples	Cryo $\delta^{13}C$	Auto $\delta^{13}C$	Cryo $\delta^{18}O$	Auto $\delta^{18}O$
Y11881	same tooth	-10.8	-11.5	-6	-6
Y19810	same tooth	-8.7	-9.2	-3.5	-2.6
Y49548	same individual	-7.8	-7.8 to -5	-3.3	-3.2 to .6
Y50803	same individual	-6.5	-7.6 to -5.4	-5	-2.7 to .4

Comparison of cryogenic and automated carbonate microsampler methods used to sample horse teeth. The microsample method was used only for tooth profiles in which teeth were sampled at least twelve times down the length of the tooth. When the same tooth was sampled by both methods, the average for the same level along the length of the tooth is reported for the microsample profiles as for the one cryogenic sample. When different teeth from the same individual are sampled, the range for the microsample profiles is reported.

Table 3.2

Isotopic values for taxa in the U-level and Dhok Pathan level

Sample No.	Locality	Taxon	Age	$\delta^{13}C$ (PDB)	$\delta^{18}O$ (PDB)
Y47024	Y317	*Sivapithecus*	9.3-9.2	-12.4	-0.9
Y13933	Y260	*Sivapithecus*	9.3-9.2	-13.1	-1.9
Y47023	Y182	*Sivapithecus*	9.3-9.2	-12.8	-1.7
Y47021	Y182	*Sivapithecus*	9.3-9.2	-13.9	-6.0
Y13811	Y224	*Sivapithecus*	9.3-9.2	-12.7	-2.7
Y51694	Y545	*Rhinoceros*	8.1-8.0	-10.9	-3.3
Y51551	Y542	*Rhinoceros*	8.1-8.0	-11.7	-3.6
Y51734	Y889	*Rhinoceros*	8.1-8.0	-13.3	-4.3
Y51572	KL11A	*Rhinoceros*	8.1-8.0	-11.4	-3.2
Y51776	Y1000	*Rhinoceros*	8.1-8.0	-8.3	-3.4
SN2	Y260	Gomphothere	9.3-9.2	-12.0	-5.5
Y5290	Y211	Gomphothere	9.3-9.2	-12.0	-5.8
Y11022	Y313	Gomphothere	9.3-9.2	-11.6	-6.5
SN1	KL01	Gomphothere	9.3-9.2	-10.1	-4.1
Y51576	Y539	Gomphothere	8.1-8.0	-11.3	-5.9
Y51774	Y1000	Gomphothere	8.1-8.0	-12.0	-1.0
Y51559	KL11A	Gomphothere	8.1-8.0	-11.0	-7.0
Y51782	Y547	Gomphothere	8.1-8.0	-12.2	-6.6
Y51704	Y545	Gomphothere	8.1-8.0	-9.6	-3.9
Y16986	Y545	*Deinotherium*	8.1-8.0	-11.7	0.0
Y3548	Y163	*B. megacephalum*	9.3-9.2	-14.6	-2.0
Y16759	KL03	*B. megacephalum*	9.3-9.2	-13.2	-2.0
Y10296	Y309	*B. megacephalum*	9.3-9.2	-10.7	-1.9
Y40145	Y403	*B. megacephalum*	9.3-9.2	-10.2	-2.0
Y11678	Y317	*B. megacephalum*	9.3-9.2	-10.2	-3.2
Y5694	Y017	*B. megacephalum*	8.1-8.0	-14.2	1.3
Y18385	Y605	*B. megacephalum*	8.1-8.0	-10.7	-0.3
Y2798	Y019	*B. megacephalum*	8.1-8.0	-11.1	-0.2
Y18412	Y605	*B. megacephalum*	8.1-8.0	-10.9	-0.9
Y14502	Y269	*P. hysudricus*	9.3-9.2	-11.0	-6.9
Y41368	Y269	*P. hysudricus*	9.3-9.2	-12.9	-9.2
Y4554	Y182	*P. hysudricus*	9.3-9.2	-12.6	-5.4
Y6753	Y227	*P. hysudricus*	9.3-9.2	-11.0	-5.7
Y10224	Y309	*P. hysudricus*	9.3-9.2	-12.0	-8.4
Y11007	Y314	*P. hysudricus*	9.3-9.2	-11.2	-8.2
Y351	Y024	*P. hysudricus*	8.1-8.0	-10.5	-5.0
Y51554	Y542	*P. hysudricus*	8.1-8.0	-11.8	-4.0
Y51700	Y545	*P. hysudricus*	8.1-8.0	-11.8	-5.9
Y49515	Y599	*P. hysudricus*	8.1-8.0	-10.2	-3.6
Y31474	Y545	*P. hysudricus*	8.1-8.0	-11.0	-2.7

Table 3.2 Continued

Isotopic values for taxa in the U-level and Dhok Pathan level

Sample No.	Locality	Taxon	Age	$\delta^{13}C$ (PDB)	$\delta^{18}O$ (PDB)
Y31474	Y545	*P. hysudricus*	8.1-8.0	-11.0	-2.7
Y13524	Y239	*H. sivalense*	9.3-9.2	-10.8	-7.3
Y47176	Y159	*H. sivalense*	9.3-9.2	-11.4	-6.6
Y12708	Y260	*H. sivalense*	9.3-9.2	-11.7	-4.5
Y5276	Y211	*H. sivalense*	9.3-9.2	-9.4	-7.0
Y4226	Y182	*H. sivalense*	9.3-9.2	-10.8	-5.1
Y27915	Y260	*H. sivalense*	9.3-9.2	-11.2	-5.7
Y51745	Y599	*H. sivalense*	8.1-8.0	-10.8	3.5
Y330	Y017	*H. sivalense*	8.1-8.0	-11.1	-4.1
Y17069	Y315	*T. magnus*	9.3-9.2	-12.8	-3.4
Y4630	Y182	*S. gandakasensis*	9.3-9.2	-13.0	-5.1
Y14499	Y269	*S. gandakasensis*	9.3-9.2	-13.1	-9.2
Y5124	Y211	*M. punjabiense*	9.3-9.2	-11.5	-7.2
Y9799	Y269	*M. punjabiense*	9.3-9.2	-12.3	-5.8
Y11797	Y317	*M. punjabiense*	9.3-9.2	-12.1	-6.9
Y47066	Y317	*Hemimeryx* lg. Sp	9.3-9.2	-8.5	-5.8
Y27044	Y224	*Dorcabune nagrii*	9.3-9.2	-11.0	-6.5
Y6857	Y227	*Dorcabune nagrii*	9.3-9.2	-10.6	-2.4
Y11155	Y310	*Dorcabune nagrii*	9.3-9.2	-9.9	-1.6
Y10220	Y309	*Dorcabune nagrii*	9.3-9.2	-11.1	-8.2
Y10886	Y312	*Dorcatherium majus*	9.3-9.2	-10.6	-2.2
Y10965	Y314	*Dorcatherium majus*	9.3-9.2	-10.5	-3.6
Y7310	Y227	*Dorcatherium majus*	9.3-9.2	-12.7	0.5
Y50921	Y227	*Dorcatherium majus*	9.3-9.2	-10.6	-8.3
Y5567	Y310	*Dorcatherium majus*	9.3-9.2	-11.1	-5.6
Y499	Y024	*Dorcatherium majus*	8.1-8.0	-10.5	-3.8
Y16270	Y539	*Dorcatherium majus*	8.1-8.0	-8.8	-2.5
Y16236	Y540	*Dorcatherium majus*	8.1-8.0	-10.6	-1.7
Y49530	Y888	*Dorcatherium majus*	8.1-8.0	-9.9	-2.4
Y679	Y033	*Dorcatherium majus*	8.1-8.0	-10.5	-2.4
Y19770	Y240	Chalicothere	9.3-9.2	-11.0	-3.3
Y19802	Y182	Chalicothere	9.3-9.2	-11.1	-3.8
Y5127-M2	Y211	*Tragoceridus*	9.3-9.2	-11.6	-6.7
Y5127-M3	Y211	*Tragoceridus*	9.3-9.2	-11.5	-3.5
Y15118-M3	Y227	*Tragoceridus*	9.3-9.2	-11.7	0.3
Y11593-M2	Y227	*Tragoceridus*	9.3-9.2	-11.7	-4.4
Y11593-M3	Y312	*Tragoceridus*	9.3-9.2	-10.7	0.7
Y9203-M2	Y017	*Tragoceridus*	8.1-8.0	-11.0	-7.0
Y9203-M3	Y017	*Tragoceridus*	8.1-8.0	-10.0	0.5
Y717011	Y545	*Tragoceridus*	8.1-8.0	-12.4	-6.5

Table 3.2 Continued

Isotopic values for taxa in the U-level and Dhok Pathan level

Sample No.	Locality	Taxon	Age	$\delta^{13}C$ (PDB)	$\delta^{18}O$ (PDB)
Y51687	Y545	*Tragoceridus*	8.1-8.0	-10.1	0.9
Y263-M2	Y017	*Tragoceridus*	8.1-8.0	-11.1	-5.1
Y263-M3	Y017	*Tragoceridus*	8.1-8.0	-9.8	-1.3
Y17007	Y545	*Tragoceridus*	8.1-8.0	-12.0	-1.9
Y2817-M2	Y017	*D. porrecticornis*	8.1-8.0	-10.9	1.6
Y2817-M3	Y017	*D. porrecticornis*	8.1-8.0	-9.6	-3.6
Y461-M2	Y024	*D. porrecticornis*	8.1-8.0	-11.9	-3.2
Y461-M3	Y024	*D. porrecticornis*	8.1-8.0	-10.7	-8.0
Y9207-M2	Y017	*D. porrecticornis*	8.1-8.0	-10.3	-1.1
Y9207-M3	Y017	*D. porrecticornis*	8.1-8.0	-10.6	-4.3
Y49550	Y606	*D. porrecticornis*	8.1-8.0	-9.3	-3.2
Y19813	Y312	*Hipparion*	9.3-9.2	-9.5	-1.5
Y47080	Y312	*Hipparion*	9.3-9.2	-8.1	-2.6
Y19810*	Y312	*Hipparion*	9.3-9.2	-9.2	-1.6
Y19810	Y312	*Hipparion*	9.3-9.2	-8.7	-3.5
Y46248	Y418	*Hipparion*	9.3-9.2	-9.1	-4.9
Y49674	Y542	*Hipparion*	8.1-8.0	-5.3	-1.3
Y18418	Y606	*Hipparion*	8.1-8.0	-7.7	-1.9
Y17859	Y599	*Hipparion*	8.1-8.0	-5.9	-2.7
Y49548*	Y890	*Hipparion*	8.1-8.0	-6.5	-0.8
Y49548	Y890	*Hipparion*	8.1-8.0	-7.8	-3.3

Carbon and oxygen isotopic values for taxa in the S-level (9.3-9.2 Ma) and the PS-level (8.1-8.0 Ma).

vegetation mosaic. Furthermore, some taxa are isotopically distinctive from others (see Tables 3.3, 3.4; Figure 3.2). These results are a good indication that true dietary signals are being preserved, for diagenetic alteration would have led to isotopic homogeneity rather than clustering by taxa. *Sivapithecus* yields some of the most depleted $\delta^{13}C$ values for the S-level, with values around −13‰, suggesting forest habitat, accompanied by some of the most enriched oxygen values (around −2.6‰), suggesting that it obtained most of its water from enriched upper canopy vegetation, most likely including fruit, as suggested by microwear (see Chapter Two). *Schizochoerus* and *Tetraconodon*, both suids, also yield $\delta^{13}C$ values around −13‰, suggesting forest habitat for both, and *Schizochoerus* also has the most depleted oxygen value, suggesting feeding in closed, wet forest. Other notable habitat preferences include the anthracothere *Hemimeryx*. While represented by only one specimen, its carbon value suggests feeding in open habitat, while its oxygen value suggests an aquatic habitat or greater dependence on water than most other taxa. These results are compatible with hippopotamus behavior today, for they are aquatic by day and feed in open habitat at night. *Bramatherium* is also isotopically similar to other fossil and modern giraffids in enriched oxygen values, suggestive of feeding on enriched upper canopy vegetation with little dependence on watering holes (Quade et al. 1995b; Zazzo et al. 2000).

When $\delta^{13}C$ values are compared between taxa, some are significantly different from others, with many of the closed habitat, or depleted, taxa going extinct before the PS-level (see Table 3.3). For example, both *Tetraconodon* and *Schizochoerus* go extinct before the PS-level sampled (Barry 2002). When compared to the

Table 3.3

Mann-Whitney comparisons of U-level taxa carbon values

	Gomph	*Bramath*	*Propot*	*Hippop*	*Schizo*	*Microb*	*Dorcab*	*Dorcath*	*Chalic*	*Trago*	*Hippar*
Sivap	*Z*=-2.45; *p*=.014	ns	*Z*=-1.78; *p*=.076	*Z*=-2.84; *p*=.005	ns	*Z*=-2.23; *p*=.025	*Z*=-2.45; *p*=.014	*Z*=-2.40 *p*=.016	*Z*=-1.94 *p*=.053	*Z*=-2.61 *p*=.009	*Z*=-2.61 *p*=.009
Gomph		ns	ns	ns	*Z*=-1.85; *p*=.064	ns	ns	ns	ns	ns	*Z*=-2.45; *p*=.014
Bramat			ns	ns	ns	ns	ns	ns	ns	ns	*Z*=-2.61; *p*=.009
Propot				ns	*Z*=-1.94; *p*=.053	ns	ns	ns	ns	ns	*Z*=-2.61; *p*=.009
Hippo					*Z*=-2.05; *p*=.040	*Z*=-2.17; *p*=.030	ns	ns	ns	ns	*Z*=-2.68; *p*=.007
Schizo						*Z*=-1.73; *p*=.083	*Z*=-1.85; *p*=.064	*Z*=-1.94; *p*=.053	ns	*Z*=-1.94; *p*=.053	*Z*=-1.94; *p*=.053
Microb							*Z*=-2.12; *p*=.034	ns	*Z*=-1.73; *p*=.083	ns	*Z*=-2.24; *p*=.025
Dorcab								ns	ns	*Z*=-1.96; *p*=.050	*Z*=-2.45; *p*=.014
Dorcat									ns	ns	*Z*=-2.61; *p*=.009
Chalic										ns	*Z*=-1.94; *p*=.053
Trago											*Z*=-2.61; *p*=.009

Mann-Whitney pair-wise comparisons of taxa for S-level $\delta^{13}C$ values indicate signficant differences between some taxa and therefore niche differentiation.

Table 3.4

Mann-Whitney comparisons of U-level taxa oxygen values

	Gomph	*Bramath*	*Propot*	*Hippop*	*Schizo*	*Microb*	*Dorcab*	*Dorcath*	*Chalic*	*Trago*	*Hippar*
Sivap	*Z*=-1.7;; *p*=.086	ns	*Z*=-2.19; *p*=.028	*Z*=-2.35; *p*=.019	ns	*Z*=-1.94; *p*=.053	ns	ns	ns	ns	ns
Gomph		*Z*=-2.46; *p*=.014	ns	ns ns	ns ns	*Z*=-1.77; *p*=.077	ns	ns	*Z*=-1.85; *p*=.028	ns	*Z*=-2.20; *p*=.028
Bramat			*Z*=-2.62; *p*=.009	*Z*=-2.85; *p*=.004	*Z*=-1.95; *p*=.051	*Z*=-2.25; *p*=.025	ns	ns	*Z*=-1.95; *p*=.051	ns	ns
Propot				ns	ns	ns	ns	*Z*=-1.78; *p*=.076	*Z*=-1.94; *p*=.053	*Z*=-2.19; *p*=.028	*Z*=-2.61; *p*=.009
Hippop					ns	ns	ns	ns	*Z*=-2.05; *p*=.040	*Z*=-2.19; *p*=.028	*Z*=-2.68; *p*=.007
Schizo						ns	ns	ns	ns	ns	*Z*=-1.94; *p*=.053
Microb							ns	ns	*Z*=-1.73; *p*=.083	*Z*=-1.94; *p*=.053	*Z*=-2.24; *p*=.025

Mann-Whitney pair-wise comparisons of taxa for S-level $\delta^{18}O$ values demonstrate significant differences between some taxa and hence niche differentiation.

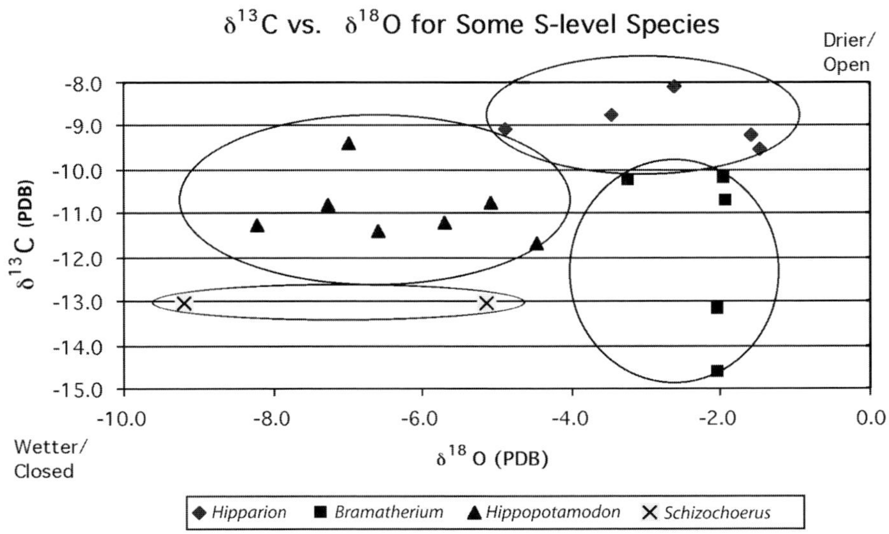

Figure 3.2 An example of specimen clustering for four S-level species, suggesting a spectrum of feeding adaptations exploiting different parts of the habitat.

other suids *Propotamochoerus* and *Hippopotamodon* in the S-level, *Schizochoerus* and *Tetraconodon* combined are significantly more depleted for $\delta^{13}C$ (average for the two is –13.0‰) (Mann-Whitney test with two-tailed *p*-value corrected for ties: $Z = -2.46$; $p = .014$). *Schizochoerus* is also more depleted than all the other taxa in the S-level that range through to the PS-level (average is –11.1‰ for all other taxa) ($Z = -2.04$; $p = .042$), as is *Sivapithecus* ($Z = -3.05$; $p = .002$). *Microbunodon*, an anthracothere that goes extinct before the PS-level, is also more depleted in $\delta^{13}C$ than most other taxa. *Dorcabune nagrii*, a tragulid that goes extinct before 8.1 Ma, does not differ significantly from *Dorcatherium majus*, a tragulid that does range through to the PS-level, but it is significantly more depleted in carbon than *Tragoceridus*, a bovid that ranges through to the PS-level. *Hipparion* is significantly more enriched than all other taxa that are represented by multiple samples, suggesting a preference for more open habitat.

There are significant differences in $\delta^{18}O$ values between taxa as well (see Table 3.4). Both *Sivapithecus* and *Bramatherium*, the giraffe, are more enriched than many other taxa, again consistent with both species obtaining much of their water from vegetation high in the upper canopy. The tragulids and bovid that did not differ significantly from *Sivapithecus* or *Bramatherium* may likewise have obtained much of their water from high-canopy fallen fruit, or the enriched values for these taxa may represent drinking in open habitat, similar to *Hipparion*. The chalicothere is also significantly more enriched than many of the other taxa. This large perissodactyl with long forelimbs and short hindlimbs may have fed and drunk in open habitats as well. Alternatively, it has been suggested that chalicotheres could stand bipedally and pull down leaves from high branches (Benton 1990), a scenario also consistent with the enriched $\delta^{18}O$ values. The suiods *Microbunodon*, *Hippopotamodon*, and *Propotamochoerus* are significantly more depleted than many other taxa, suggesting they preferred more closed habitat. Again, *Hipparion* is more enriched than many other taxa suggesting an affinity for open habitat.

In addition to sampling a wide range of fauna, this investigation also sampled many fossil localities. A total of 38 localities or collection levels were represented, 21 of which date to the S-level, with many sites represented by multiple specimens and taxa (see Tables 3.5 and 3.6). While this study was not designed to profile individual sites, some initial inferences can be made from the data collected, including inferences about *Sivapithecus* habitat. Of the S-level sites, *Sivapithecus* is found at nine of them. Among these, the six sites represented by at least three specimens all contain at least one specimen other than *Sivapithecus* with a $\delta^{13}C$ value of –12‰ or lower, suggesting the presence of closed forest, and one site (Y182) includes *Schizochoerus*, the taxon with the most depleted carbon and oxygen values thought to prefer closed, wet forest. Three of the nine sites (Y211, Y310, and Y317) also include a $\delta^{13}C$ value above –10‰, indicating some open habitat present as well. All *Sivapithecus* sites contain specimens with at least the average $\delta^{18}O$ value for the S-level (–5‰) or lower, with three of the sites (Y227, Y309, Y314) including samples with $\delta^{18}O$ values lower than –8‰, suggestive of non-evaporative, or closed, habitat. At only one S-level site sampled with oxygen values that low is no *Sivapithecus* found. That locality is Y269, and all of its taxa (all suoids) are depleted in both carbon and oxygen, and again, *Schizochoerus* is present. Y269 thus appears to be closed, wet forest, but no *Sivapithecus* is found there.

An S-level locality which is unusually enriched in both carbon and oxygen is Y312. This locality is represented by a bovid, tragulid, and four equids, and while biased towards open-habitat equids, the bovid and tragulid specimens are enriched in both carbon and oxygen compared to their averages for the S-level, and the $\delta^{18}O$ value for *Tragoceridus* is the highest value for the S-level. These results suggest that Y312 was an open habitat, and unusually open for the S-level. No *Sivapithecus* is found there.

Finally, the S-level can be compared to other Miocene hominoid sites, Pasalar and Ft. Ternan (see Table 3.7). The S-level has a greater range of $\delta^{13}C$ values than either Ft. Ternan or Pasalar, with values both more depleted and more enriched than either site, suggesting a greater mosaic of habitats including both more closed and more open patches. It should be noted that a greater range of habitats at the S-level is expected, for data represent many localities, while Ft. Ternan and Pasalar are each one locality and hence one snapshot in time. The

Table 3.5

Isotopic samples sorted by locality

Sample No.	Locality	Taxon	Age	$\delta^{13}C$ (PDB)	$\delta^{18}O$ (PDB)
SN1	KL01	Gomphothere	9.3-9.2	-10.1	-4.1
Y16759	KL03	*Bramatherium*	9.3-9.2	-13.2	-2.0
Y47176	Y159	*Hippopotamodon*	9.3-9.2	-11.4	-6.6
Y3548	Y163	*Bramatherium*	9.3-9.2	-14.6	-2.0
Y47023	**Y182**	***Sivapithecus***	**9.3-9.2**	**-12.8**	**-1.7**
Y47021	**Y182**	***Sivapithecus***	**9.3-9.2**	**-13.9**	**-6.0**
Y4630	**Y182**	***Schizochoerus***	**9.3-9.2**	**-13.0**	**-5.1**
Y4554	**Y182**	***Propotamochoerus***	**9.3-9.2**	**-12.6**	**-5.4**
Y19802	**Y182**	**Chalicothere**	**9.3-9.2**	**-11.1**	**-3.8**
Y4226	**Y182**	***Hippopotamodon***	**9.3-9.2**	**-10.8**	**-5.1**
RANGE				3.1	4.3
Y5290	**Y211**	***Gomphothere***	**9.3-9.2**	**-12.0**	**-5.8**
Y5127-M2	**Y211**	***Tragoceridus***	**9.3-9.2**	**-11.6**	**-6.7**
Y5124	**Y211**	***Microbunodon***	**9.3-9.2**	**-11.5**	**-7.2**
Y5127-M3	**Y211**	***Tragoceridus***	**9.3-9.2**	**-11.5**	**-3.5**
Y5276	**Y211**	***Hippopotamodon***	**9.3-9.2**	**-9.4**	**-7.0**
RANGE				**2.6**	**3.7**
Y13811	**Y224**	***Sivapithecus***	**9.3-9.2**	**-12.7**	**-2.7**
Y27044	**Y224**	***Dorcabune nagrii***	**9.3-9.2**	**-11.0**	**-6.5**
RANGE				1.7	3.8
Y7310	**Y227**	***Dorcatherium majus***	**9.3-9.2**	**-12.7**	**0.5**
Y15118-M3	**Y227**	***Tragoceridus***	**9.3-9.2**	**-11.7**	**0.3**
Y11593-M2	**Y227**	***Tragoceridus***	**9.3-9.2**	**-11.7**	**-4.4**
Y6753	**Y227**	***Propotamochoerus***	**9.3-9.2**	**-11.0**	**-5.7**
Y50921	**Y227**	***Dorcatherium majus***	**9.3-9.2**	**-10.6**	**-8.3**
Y6857	**Y227**	***Dorcabune nagrii***	**9.3-9.2**	**-10.6**	**-2.4**
RANGE				**2.1**	**8.8**
Y13524	Y239	*Hippopotamodon*	9.3-9.2	-10.8	-7.3
Y19770	Y240	Chalicothere	9.3-9.2	-11.0	-3.3
Y13933	**Y260**	***Sivapithecus***	**9.3-9.2**	**-13.1**	**-1.9**
SN2	**Y260**	***Gomphothere***	**9.3-9.2**	**-12.0**	**-5.5**
Y12708	**Y260**	***Hippopotamodon***	**9.3-9.2**	**-11.7**	**-4.5**
Y27915	**Y260**	***Hippopotamodon***	**9.3-9.2**	**-11.2**	**-5.7**
RANGE				**1.9**	**3.8**

Table 3.5 Continued

Isotopic samples sorted by locality

Sample No.	Locality	Taxon	Age	$\delta^{13}C$ (PDB)	$\delta^{18}O$ (PDB)
Y14499	Y269	*Schizochoerus*	9.3-9.2	-13.1	-9.2
Y41368	Y269	*Propotamochoerus*	9.3-9.2	-12.9	-9.2
Y9799	Y269	*Microbunodon*	9.3-9.2	-12.3	-5.8
Y14502	Y269	*Propotamochoerus*	9.3-9.2	-11.0	-6.9
RANGE				**2.1**	**3.4**
Y10224	**Y309**	***Propotamochoerus***	**9.3-9.2**	**-12.0**	**-8.4**
Y10220	**Y309**	***Dorcabune nagrii***	**9.3-9.2**	**-11.1**	**-8.2**
Y10296	**Y309**	***Bramatherium***	**9.3-9.2**	**-10.7**	**-1.9**
RANGE				**1.3**	**6.5**
Y5567	**Y310**	***Dorcatherium majus***	**9.3-9.2**	**-11.1**	**-5.6**
Y11155	**Y310**	***Dorcabune nagrii***	**9.3-9.2**	**-9.9**	**-1.6**
RANGE				**1.2**	**4.0**
Y11593-M3	Y312	*Tragoceridus*	9.3-9.2	-10.7	0.7
Y10886	Y312	*Dorcatherium majus*	9.3-9.2	-10.6	-2.2
Y19813	Y312	*Hipparion*	9.3-9.2	-9.5	-1.5
Y19810*	Y312	*Hipparion*	9.3-9.2	-9.2	-1.6
Y19810	Y312	*Hipparion*	9.3-9.2	-8.7	-3.5
Y47080	Y312	*Hipparion*	9.3-9.2	-8.1	-2.6
RANGE				**2.6**	**4.2**
Y11022	Y313	Gomphothere	9.3-9.2	-11.6	-6.5
Y11007	**Y314**	***Hippopotamodon***	**9.3-9.2**	**-11.2**	**-8.2**
Y10965	**Y314**	***Dorcatherium majus***	**9.3-9.2**	**-10.5**	**-3.6**
RANGE				**0.7**	**4.6**
Y17069	Y315	*Tetraconodon*	9.3-9.2	-12.8	-3.4
Y47024	**Y317**	***Sivapithecus***	**9.3-9.2**	**-12.4**	**-0.9**
Y11797	**Y317**	***Microbunodon***	**9.3-9.2**	**-12.1**	**-6.9**
Y11678	**Y317**	***Bramatherium***	**9.3-9.2**	**-10.2**	**-3.2**
Y47066	**Y317**	***Hemimeryx* lg. Sp**	**9.3-9.2**	**-8.5**	**-5.8**
RANGE				**3.9**	**6.0**
Y40145	Y403	*Bramatherium*	9.3-9.2	-10.2	-2.0
Y46248	Y418	*Hipparion*	9.3-9.2	-9.1	-4.9
Y51572	KL11A	Rhino	8.1-8.0	-11.4	-3.2
Y51559	KL11A	Gomphothere	8.1-8.0	-11.0	-7.0
RANGE				**0.4**	**3.8**

Table 3.5 Continued

Isotopic samples sorted by locality

Sample No.	Locality	Taxon	Age	$\delta^{13}C$ (PDB)	$\delta^{18}O$ (PDB)
Y51572	KL11A	Rhino	8.1-8.0	-11.4	-3.2
Y51559	KL11A	Gomphothere	8.1-8.0	-11.0	-7.0
RANGE				**0.4**	**3.8**
Y5694	Y017	*Bramatherium*	8.1-8.0	-14.2	1.3
Y263-M2	Y017	*Tragoceridus*	8.1-8.0	-11.1	-5.1
Y330	Y017	*Hippopotamodon*	8.1-8.0	-11.1	-4.1
Y9203-M2	Y017	*Tragoceridus*	8.1-8.0	-11.0	-7.0
Y2817-M2	Y017	*Dorcadoxa*	8.1-8.0	-10.9	1.6
Y9207-M3	Y017	*Dorcadoxa*	8.1-8.0	-10.6	-4.3
Y9207-M2	Y017	*Dorcadoxa*	8.1-8.0	-10.3	-1.1
Y9208	Y017	*Tragoceridus*	8.1-8.0	-10.3	0.1
Y9203-M3	Y017	*Tragoceridus*	8.1-8.0	-10.0	0.5
Y263-M3	Y017	*Tragoceridus*	8.1-8.0	-9.8	-1.3
Y2817-M3	Y017	*Dorcadoxa*	8.1-8.0	-9.6	-3.6
RANGE				**4.6**	**8.6**
Y2798	Y019	*Bramatherium*	8.1-8.0	-11.1	-0.2
Y461-M2	Y024	*Dorcadoxa*	8.1-8.0	-11.9	-3.2
Y461-M3	Y024	*Dorcadoxa*	8.1-8.0	-10.7	-8.0
Y351	Y024	*Propotamochoerus*	8.1-8.0	-10.5	-5.0
Y499	Y024	*Dorcatherium majus*	8.1-8.0	-10.5	-3.8
RANGE				**1.4**	**4.8**
Y679	Y033	*Dorcatherium majus*	8.1-8.0	-10.5	-2.4
Y51576	Y539	Gomphothere	8.1-8.0	-11.3	-5.9
Y16270	Y539	*Dorcatherium majus*	8.1-8.0	-8.8	-2.5
RANGE				**2.5**	**3.4**
Y16236	Y540	*Dorcatherium majus*	8.1-8.0	-10.6	-1.7
Y51554	Y542	*Propotamochoerus*	8.1-8.0	-11.8	-4.0
Y51551	Y542	Rhino	8.1-8.0	-11.7	-3.6
Y49674	Y542	*Hipparion*	8.1-8.0	-5.3	-1.3
RANGE				**6.5**	**2.7**
Y717011	Y545	*Tragoceridus*	8.1-8.0	-12.4	-6.5
Y17007	Y545	*Tragoceridus*	8.1-8.0	-12.0	-1.9
Y51700	Y545	*Propotamochoerus*	8.1-8.0	-11.8	-5.9
Y16986	Y545	*Deinotherium*	8.1-8.0	-11.7	0.0
Y31474	Y545	*Propotamochoerus*	8.1-8.0	-11.0	-2.7

Table 3.5 Continued

Isotopic samples sorted by locality

Sample No.	Locality	Taxon	Age	$\delta^{13}C$ (PDB)	$\delta^{18}O$ (PDB)
Y51694	Y545	Rhino	8.1-8.0	-10.9	-3.3
Y51687	Y545	*Tragoceridus*	8.1-8.0	-10.1	0.9
Y51704	Y545	*Gomphothere*	8.1-8.0	-9.6	-3.9
RANGE				**2.8**	**6.5**
Y51782	Y547	Gomphothere	8.1-8.0	-12.2	-6.6
Y51745	Y599	*Hippopotamodon*	8.1-8.0	-10.8	3.5
Y49515	Y599	*Propotamochoerus*	8.1-8.0	-10.2	-3.6
Y17859	Y599	*Hipparion*	8.1-8.0	-5.9	-2.7
RANGE				**4.9**	**7.1**
Y18412	Y605	*Bramatherium*	8.1-8.0	-10.9	-0.9
Y18385	Y605	*Bramatherium*	8.1-8.0	-10.7	-0.3
RANGE				**0.2**	**0.6**
Y49550	Y606	*Dorcadoxa*	8.1-8.0	-9.3	-3.2
Y18418	Y606	*Hipparion*	8.1-8.0	-7.7	-1.9
RANGE				**1.6**	**1.3**
Y49530	Y888	*Dorcatherium majus*	8.1-8.0	-9.9	-2.4
Y51734	Y889	Rhino	8.1-8.0	-13.3	-4.3
Y49548	Y890	*Hipparion*	8.1-8.0	-7.8	-3.3
Y49548*	Y890	*Hipparion*	8.1-8.0	-6.5	-0.8
RANGE				**1.3**	**2.5**
Y51774	Y1000	Gomphothere	8.1-8.0	-12.0	-1.0
Y51776	Y1000	Rhino	8.1-8.0	-8.3	-3.4
RANGE				**3.7**	**2.4**

Isotopic samples sorted by locality. *Sivapithecus* sites in bold.

Table 3.6

Summary of carbon and oxygen values for localities

Locality	Number samples	Number taxa	Age (Ma)	Lowest $\delta^{13}C$	Highest $\delta^{13}C$	Range $\delta^{13}C$	Lowest $\delta^{13}C$	Highest $\delta^{18}O$	Range $\delta^{18}O$
Y224	**2**	**2**	**9.3-9.2**	**-12.7**	**-11.0**	**1.7**	**-6.5**	**-2.7**	**3.8**
Y310	**2**	**2**	**9.3-9.2**	**-11.1**	**-9.9**	**1.2**	**-5.6**	**-1.6**	**4.0**
Y314	**2**	**2**	**9.3-9.2**	**-11.2**	**-10.5**	**0.7**	**-8.2**	**-3.6**	**4.6**
Y260	**4**	**3**	**9.3-9.2**	**-13.1**	**-11.2**	**1.9**	**-5.7**	**-1.9**	**3.8**
Y309	**3**	**3**	**9.3-9.2**	**-12**	**-10.7**	**1.3**	**-8.4**	**-1.9**	**6.5**
Y317	**4**	**4**	**9.3-9.2**	**-12.4**	**-8.5**	**3.9**	**-6.9**	**-0.9**	**6.0**
Y182	**6**	**5**	**9.3-9.2**	**-13.9**	**-10.8**	**3.1**	**-6.0**	**-1.7**	**4.3**
Y269	4	3	9.3-9.2	-13.1	-11	2.1	-9.2	-5.8	3.4
Y211	**5**	**4**	**9.3-9.2**	**-12**	**-9.4**	**2.6**	**-7.2**	**-3.5**	**3.7**
Y227	**6**	**4**	**9.3-9.2**	**-12.7**	**-10.6**	**2.1**	**-8.3**	**0.5**	**8.8**
Y312	6	3	9.3-9.2	-10.7	-8.1	2.6	-3.5	0.7	4.2
Y1000	2	2	8.1-8.0	-12	-8.3	3.7	-3.4	-1	2.4
Y539	2	2	8.1-8.0	-11.3	-8.8	2.5	-5.9	-2.5	3.4
Y605	2	1	8.1-8.0	-10.9	-10.7	0.2	-0.9	-0.3	0.6
Y606	2	2	8.1-8.0	-9.3	-7.7	1.6	-3.2	-1.9	1.3
Y890	2	1	8.1-8.0	-7.8	-6.5	1.3	-3.3	-0.8	2.5
Y542	3	3	8.1-8.0	-11.8	-5.3	6.5	-4	-1.3	2.7
Y599	3	3	8.1-8.0	-10.8	-5.9	4.9	-3.6	3.5	7.1
Y024	4	3	8.1-8.0	-11.9	-10.5	1.4	-8	-3.2	4.8
Y545	7	5	8.1-8.0	-12.4	-9.6	2.8	-6.5	0	6.5
Y017	11	4	8.1-8.0	-14.2	-9.6	4.6	-7	1.6	8.6

Summary of $\delta^{13}C$ and $\delta^{18}O$ values for localities represented by multiple samples. *Sivapithecus* sites in bold.

Table 3.7

Comparison of Miocene hominoid localities

	Ft. Ternan	Pasalar	S-level
$\delta^{13}C$ values	-13.0 to -8.6	-13.5 to -9.0	-14.6 to -8.1
Range $\delta^{13}C$	4.4	4.5	6.5
Average $\delta^{13}C$	-10.8	-11.2	-11.3
$\delta^{18}O$ values	-.7 to +4.2	-9.1 to +1.4	-9.2 to +.7
Range $\delta^{18}O$	4.9	10.5	9.9
Average $\delta^{18}O$	1.5	-4.6	-4.5

Comparison of Miocene hominoid localities, including Ft. Ternan, represented by 21 specimens and 9 species; Pasalar, represented by 44 specimens and 16 species; and the S-level localities, represented by 54 specimens and 14 species. Ft. Ternan data from Cerling et al. (1997). Pasalar data from Quade et al. (1995). All data relative to the standard PDB.

average $\delta^{13}C$ value for each of the three regions is the same, suggesting that all had similar forest and woodland habitats. The S-level and Pasalar are also very similar in their oxygen values, with similar ranges, suggesting similarities in percentage closed versus open habitat, in degree of seasonality, or possibly in source of precipitation as well. Ft. Ternan, however, yields enriched $\delta^{18}O$ values compared to Pasalar and the S-level, indicating a different rain source, and its range of values is only half that of the other localities. While this smaller range could be due to a smaller and less diverse faunal sample, given the carbon ranges are comparable to the other localities, a reduced range in $\delta^{18}O$ may indicate a less variable and possibly drier climate.

Post-*Sivapithecus* (PS)-level (8.1–8.0 Ma)

Carbon results for the PS-level fall between −14.2 and −5.3‰ (in contrast to −14.6 to −8.1‰ for the S-level), for a range of 8.9‰ (versus 6.5‰) and a mean of −10.5‰ (versus −11.3‰) (see Table 3.2 and Figure 3.3). *Hipparion* clearly is eating some C_4 grass, while all other taxa, including bovids, are still C_3 browsers. The greater range of values suggests a more varied mosaic of habitat, with closed woodland or forest still present, but with some C_4 grasses as well. The fact that bovids, including *Dorcadoxa*, a reduncine (all grazers today), had not incorporated C_4 graze into their diets by 8.0 Ma suggests that while C_4 patches were present, they were not extensive.

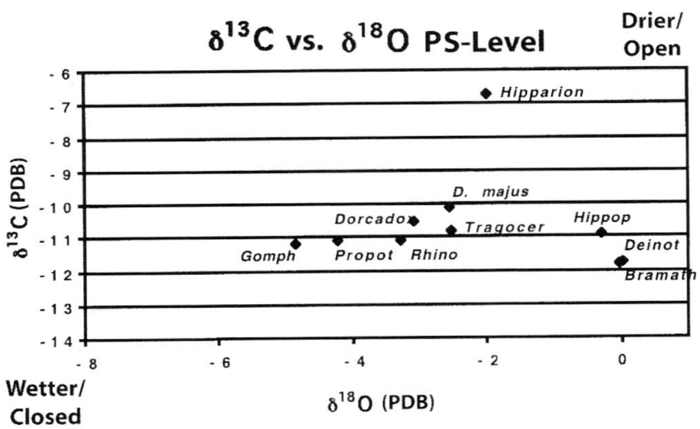

Figure 3.3 Average carbon and oxygen isotope values for each taxon in the PS- level indicate a range of feeding adaptations exploiting different parts of the habitat. Taxa include gomphotheres and deinotheres; rhinoceros; the suids *Propotamochoerus* and *Hippopotamodon*; the tragulid *Dorcatherium majus*; the bovids *Tragoceridus* and *Dorcadoxa*; the giraffid *Bramatherium*; and the equid *Hipparion*.

The $\delta^{18}O$ values for the PS-level fall between −8.0 and +3.5‰ (compared to −9.2 to +.7‰ for the S–level), for a range of 11.5‰ (versus 9.9‰) and an average of −2.7‰ (versus −4.5‰). Again, for several taxa, individuals cluster in their $\delta^{13}C$ and/or $\delta^{18}O$ values, suggesting they fed in particular parts of the habitat, and some taxa are isotopically distinct from others (see Table 3.8, 3.9; Figure 3.4). *Bramatherium* and gomphotheres continue to yield depleted carbon values indicative of forest, and they are significantly more depleted in $\delta^{13}C$ values than several other PS-level taxa, suggesting the number of forest taxa has decreased since the S-level. *Dorcadoxa* is the only species sampled which was not present during the S-level, and its average $\delta^{13}C$ values of −10.4‰ suggests this new species had an affinity for open woodland. *Hipparion* is again significantly more enriched than any other taxon represented by more than one specimen, indicating that it fed in open habitat more often than other taxa.

Taxa cluster in oxygen values as well. *Bramatherium* remains more enriched in $\delta^{18}O$ values than most other taxa, again suggesting feeding from the enriched upper canopy. *Propotamochoerus* remains more depleted than some taxa, suggesting that it still preferred more closed, less evaportive habitats. *Hipparion* is also still more enriched than several taxa, also indicative of open habitat.

Comparison of the S-level and the PS-level

When all taxa are combined per level, the PS-level carbon values are significantly more enriched ($Z = -2.64$, $p = .008$). When horses, the only substantial C_4 grazers, are removed from both levels (resulting in a range of −14.6 to −8.5‰ for the S-level and −14.2 to −8.3‰ for the PS-level), this difference is still significant ($Z = -2.79$, $p = .005$). When $\delta^{18}O$ values are combined for all taxa for each level, the PS-level taxa are again significantly enriched with respect to those from the S-level ($Z = -3.09$; $p = .002$). Both the carbon and oxygen isotopes therefore indicate a drier, more open habitat with a changing climate by 8.1 Ma (see Figure 3.5).

Table 3.8

Mann-Whitney comparisons of Dhok Pathan level taxa carbon values

	Rhino	*Gomph*	*Bramath*	*Propot*	*Hippop*	*D. majus*	*Tragocer*	*Dorcadox*	*Hipparion*
Rhino		ns	ns	ns	ns	ns	ns	ns	*Z*=-2.61; *p*=.009
Gomph			ns	ns	ns	*Z*=-1.78; *p*=.075	ns	ns	*Z*=-2.61; *p*=.009
Bramath				ns	ns	*Z*=-2.46; *p*=.014	ns ns	*Z*=-1.71; *p*=.088	*Z*=-2.45; *p*=.014
Propot					ns	ns	ns	ns	*Z*-2.62; *p*=.009
Hippop						*Z*=-1.95; *p*=.051	ns	ns	*Z*=-1.94; *p*=.053
D. majus							ns	ns	*Z*=-2.62; *p*=.009
Tragocer								ns	*Z*=-2.93; *p*=.003
Dorcadox									*Z*=-2.84; *p*=.005

Mann-Whitney pair-wise comparisons of taxa for PS-level $\delta^{13}C$ values indicate significant differences between some taxa and hence niche differentiation.

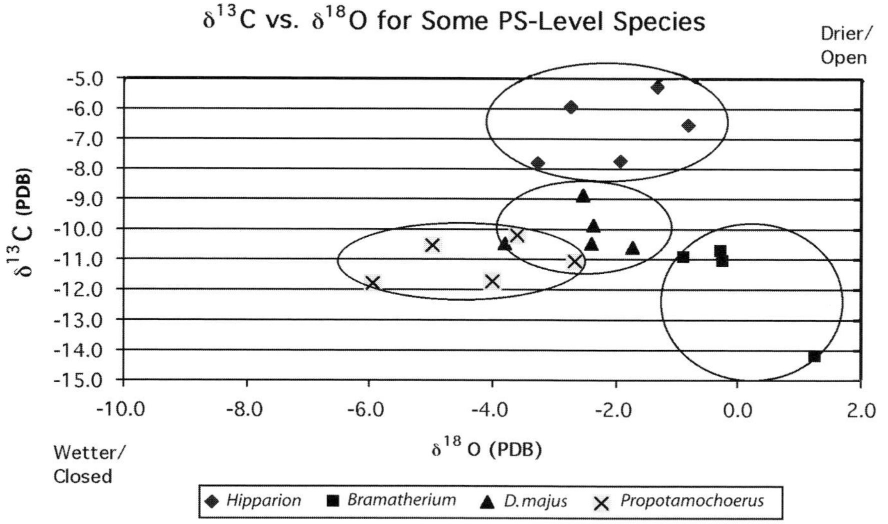

Figure 3.4 An example of specimen clustering for four PS-level species, suggesting a spectrum of feeding adaptations exploiting different parts of the habitat.

Table 3.9

Mann-Whitney comparisons of Dhok Pathan level taxa oxygen values

	Rhino	*Gomph*	*Bramath*	*Propot*	*Hippop*	*D. majus*	*Tragocer*	*Dorcadox*	*Hipparion*
Rhino		ns	Z=-2.45; p=.014	ns	ns	Z=-1.78; p=.075	ns	ns	Z=-2.30; p=.021
Gomph			Z=-2.45; p=.014	ns ns	ns	ns	ns	ns	Z=-1.78; p=.076
Bramath				Z=-2.45; p=.014	ns	Z=-2.46; p=.014	ns	Z=-1.89; p=.058	Z=-2.20; p=.028
Propot					ns	Z=-2.20; p=.028	ns	ns	Z=-2.30; p=.021
Hippop						ns	ns	ns	ns
D. majus							ns	ns	ns
Tragocer								ns	ns
Dorcadox									ns

Mann-Whitney pair-wise comparisons of taxa for PS-level $\delta^{18}O$ values demonstrate significant differences between some taxa and therefore niche differentiation.

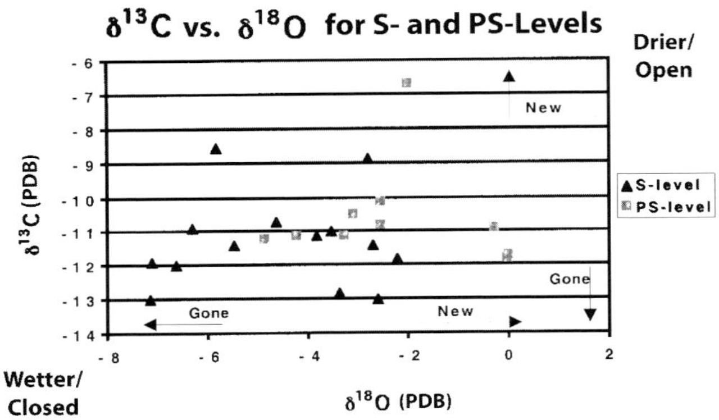

Figure 3.5 Average values for each taxon in the S-level and the PS-level. A shift to more positive $\delta^{13}C$ and $\delta^{18}O$ values over time suggests a decrease in forest habitat, an increase in open habitat, and possibly a shift in rainfall regime.

The enrichment of 1.8‰ in $\delta^{18}O$ between the different levels, if used as an estimate for change in meteoric water, can be interpreted as a change in temperature using an equation by Dansgaard (1964):

$\delta^{18}O_{\text{meteoric water}} = .69T - 13.6‰$ where T = mean annual air temperature.

Using this equation, the 2.3‰ enrichment translates into about a three degrees Celsius increase in mean annual temperature, if temperature were the only climatic factor to change. Likewise, if change in precipitation were the only factor, then at tropical latitudes where there is a linear relationship between average monthly rainfall and its mean monthly $\delta^{18}O$ value, about a −1.5‰ shift per 100 mm, then the 1.8‰ enrichment translates into about a 120 mm reduction in rainfall.

These temperature and precipitation changes, when compared to climatic data from evergreen, semi-evergreen, moist-, and dry-monsoon forests of India today, are not enough to account for major changes in vegetation. For example, there is a three degress Celsius spread between evergreen and semi-evergreen forests, but monsoonal forests fall between these values, making any interpretation of small temperature changes ambiguous. Amount of precipitation varies considerably more between the modern ecosystems, however, with a reduction of over 1,000 mm between evergreen and semi-evergreen, a further reduction of over 400 mm from semi-evergreen to moist-monsoonal, and another reduction of 250 mm from moist- to dry-monsoonal. Furthermore, for any of these ecosystem types, precipitation values can have 500 to over 1,000 mm ranges (Walter 1973).

Thus a reduction of 120 mm is not substantial enough to cause a major shift in vegetation unless accompanied by other climatic or atmospheric changes. The most important single determining factor of the modern forest types is length of the dry period, or seasonality of rainfall. The oxygen enrichment of the Siwaliks cannot be interpreted solely as a shift to summer rains, for heavy summer rains of monsoons are greatly depleted relative to winter rains due to the amount effect mentioned earlier. However, an increase in seasonality, when combined with other factors such as decrease in rainfall or change in precipitation source, could have accounted for both the carbon and oxygen shifts observed. Changes in the atmosphere, such as a decrease of atmospheric CO_2 levels could have also played a role, and this hypothesis will be explored further in Chapter Five.

An alternative explanation is that the decrease in rainfall was more substantial than detected. If summer versus winter rains differed greatly in $\delta^{18}O$ values, as in monsoonal regions today, some of the variability observed in oxygen values is likely the result of sampling different seasons of the year for different specimens. This seasonal variability may have confounded differences in amount of rainfall if samples from the different time levels were skewed to different seasons. More detailed reconstructions of rainfall regimes which allow comparison of particular seasons are presented in Chapter Four.

In addition to changes in the vegetation mosaic and climate, changes in feeding behavior or habitat preferences can be studied for individual taxa. For seven of the species studied, specimens were available for both stratigraphic levels. When average $\delta^{13}C$ values are compared for each taxon between the two levels, taxa become either enriched or have the same value from the S-level to the PS-level. For *Dorcatherium majus* and *Hipparion*, the enrichment is significant (see Table 3.10), suggesting a shift in feeding to more open habitats. When average $\delta^{18}O$ values are likewise compared for the seven taxa, all show enrichment from the S-level to the PS-level, with significantly enriched values for *Bramatherium* and the two suids *Propotamochoerus* and *Hippopotamodon*, suggesting a shift to more open habitat or a shift in climate (see Table 3.11).

The range of isotopic values, and hence niche breadth, and changes in breadth over time can also be evaluated per taxon (see Table 3.12). *Bramatherium* exhibits a wide range of carbon values in both levels, indicative of feeding throughout the forest canopy, while its oxygen range is relatively small, suggesting that it obtained most of its water from enriched upper canopy leaves rather than watering holes. Rhinos also show a wide range of carbon values but a small range of oxygen values. Modern rhinos are highly omnivorous, feeding on fruits, leaves, branches, shrubs, and grass in a range of habitats (Nowak 1999; Kingdon 1997). Miocene rhino carbon values are compatible with the modern analogue. Furthermore, modern rhinos depend upon bodies of water for temperature control, digestion, and

Table 3.10

Comparison of carbon values for taxa in both U- and Dhok Pathan level

Taxon	$\delta^{13}C$ S-level	$\delta^{13}C$ PS-level	Mann Whitney
Gomphothere	-11.4 (4)	-11.2 (5)	ns
Bramatherium	-11.8 (5)	-11.8 (4)	ns
Propotamochoerus	-11.8 (6)	-11.1 (5)	ns
Hippopotamodon	-10.9 (6)	-10.9 (2)	ns
Dorcatherium majus	-11.1 (5)	-10.1 (5)	$Z=-2.03$; $p=.042$
Tragoceridus	-11.4 (5)	-10.8 (8)	ns
Hipparion	-8.9 (5)	-6.7 (5)	$Z=-2.45$; $p=.014$

Comparison of average $\delta^{13}C$ values for taxa sampled in both the S-level (9.3-9.2 Ma) and the PS-level (8.1-8.0 Ma). Number of samples in parentheses. For all taxa, carbon values either do not change or they become enriched over time.

ns=not significant

Table 3.11

Comparison of oxygen for taxa in both the U- and Dhok Pathan level

Taxon	$\delta^{18}O$ S-level	$\delta^{18}O$ PS-level	Mann Whitney
Gomphothere	-5.5 (4)	-4.9 (5)	ns
Bramatherium	-2.2 (5)	-0.033 (4)	Z=-2.49; p=.013
Propotamochoerus	-7.3 (6)	-4.2 (5)	Z=-2.37; p=.018
Hippopotamodon	-6.0 (6)	-0.3 (2)	Z=-2.00; p=.046
Dorcatherium majus	-3.9 (5)	-2.6 (5)	ns
Tragoceridus	-2.7 (5)	-2.5 (8)	ns
Hipparion	-2.8 (5)	-2 (5)	ns

Comparison of average $\delta^{18}O$ values for taxa sampled in both the S-level (9.3-9.2 Ma) and PS-level (8.1-8.0 Ma). Number of samples in parentheses. For all taxa, oxygen values become enriched over time.

ns=not significant

scent communication (Kingdon 1997). The depleted Siwalik oxygen results are compatible with water dependence, and the small range of values are compatible with dependence on large bodies of water that would have been buffered from seasonal changes in $\delta^{18}O$ values. Notable taxa with large ranges in oxygen include both tragulids and bovids in the S-level and bovids in the PS-level, suggesting they are obtaining water from more varied sources than other taxa, either due to use of both open and closed habitats, or to obtaining some forest floor food as well as fruits that fell from the upper canopy.

Some taxa show a change in range of their carbon or oxygen values, suggesting a shift in niche. *Tragoceridus* and *Hipparion* show an increase in breadth of carbon values, with the PS-level range being more

Table 3.12

Range of isotopic values per taxon in both U- and Dhok Pathan level

Taxon	Range $\delta^{13}C$ S	Range $\delta^{13}C$ PS	Range $\delta^{18}O$ S	Range $\delta^{18}O$ PS
Rhino	NA	5(5)	NA	1.1 (5)
Gomphothere	1.9 (4)	2.6 (5)	2.4 (4)	6 (5)
Bramatherium	4.4 (5)	3.5 (4)	1.3 (5)	1.1 (4)
Propotamochoerus	1.9 (6)	1.6 (5)	3.8 (6)	3.2 (5)
Hippopotamodon	2.3 (6)	0.3 (2)	2.8 (6)	7.6 (2)
Schizochoerus	0.1 (2)	NA	4.1 (2)	NA
Microbunodon	0.8 (3)	NA	1.4 (3)	NA
Dorcabune nagrii	1.2 (4)	NA	6.6 (4)	NA
Dorcatherium majus	2.2 (5)	1.8 (5)	8.8 (5)	2.1 (5)
Chalicothere	0.1 (2)	NA	0.5 (2)	NA
Tragoceridus	1 (5)	2.6 (8)	7.4 (5)	7.5 (8)
Dorcadoxa	NA	2.6 (7)	NA	9.6 (7)
Hipparion	1.4 (5)	2.5 (5)	3.4 (5)	2.5 (5)

Range (per mil) of $\delta^{13}C$ and $\delta^{18}O$ values per taxon for both the S-level (S) and the PS-level (PS). Number of specimens given in parentheses.

than a per mil greater than the S-level samples. This increase might indicate a slightly greater use of more varied habitats. It should also be noted that *Dorcadoxa*, the PS-level bovid that was not present during the S-level, shows the greatest range in oxygen values (9.6‰) compared to any other taxa in either level, suggesting that this new species exploited a range of habitats including both closed and open. If *Dorcadoxa* and the increased niche breadth inferred for PS-level *Tragoceridus* and *Hipparion* are representative of fauna in general for that time level, it suggests that in this time of climatic change and a more varied vegetation mosaic, fauna are becoming more generalized in their habitat preferences. The one taxon which showed a drastic reduction in its oxygen range in the PS-level is *D. majus*, the only tragulid sampled still present from the S-level. This decrease suggests a reduction in niche breadth, perhaps due to competition with other taxa such as bovids, or perhaps due to a loss of preferred habitats.

Finally, changes in range of values per locality suggest a more varied vegetation mosaic on a local scale. Several PS-level localities show a wider range in $\delta^{13}C$ values compared to the S-level (see Figure 3.6). While this difference in combined ranges of carbon values is not statistically significant between the S-level and PS-level, only two of 11 S-level localities has a range in $\delta^{13}C$ over 3‰, while four out of ten PS-level localities have ranges over 3‰, with three having large ranges of 4.6, 4.9, and 6.5‰. These greater ranges reflect a more varied mosaic of local habitats, with two of these localities (Y542 and Y549) showing clear evidence of C_4 grasslands in conjunction with still fairly closed habitats. Furthermore, those sites with the largest ranges were also the ones with the highest carbon values, also indicating that C_4 grasses are found in conjunction with more closed habitats rather than forming extensive grasslands (see Figure 3.7). Furthermore, while C_4 patches of grass make their appearance in the PS-level, localities of relatively closed habitat with no open patches are still present (see Figure 3.8).

While PS-level localities exhibit greater ranges in carbon values, some S-level localities as well as some PS-level localities exhibit high ranges in oxygen values, suggesting a similar spectrum of evaporative environments (see Figure 3.9). S-level localities do not exhibit the

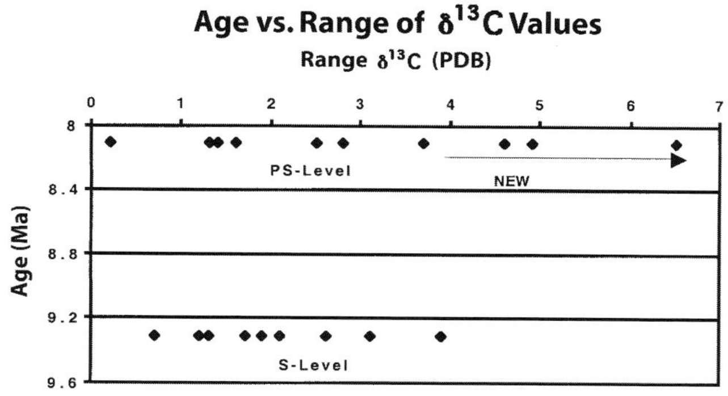

Figure 3.6 Range in $\delta^{13}C$ values for each S-level and PS-level locality. Several PS-level localities exhibit larger ranges in carbon values than S-level localities, suggesting more variable habitats at 8.1-8.0 Ma than at 9.3-9.2 Ma.

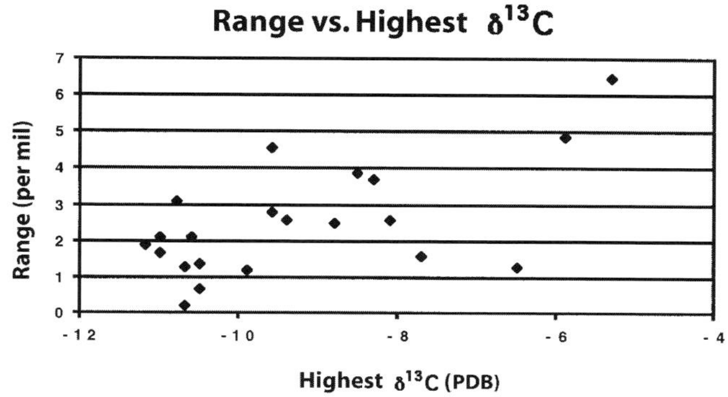

Figure 3.7 Range of $\delta^{13}C$ values vs. highest $\delta^{13}C$ values for each locality in both the S-level and the PS-level. Localities with carbon values indicative of some C_4 grasses show the widest ranges, suggesting that C_4 grasses are often found in conjunction with more closed habitat rather than as extensive grasslands.

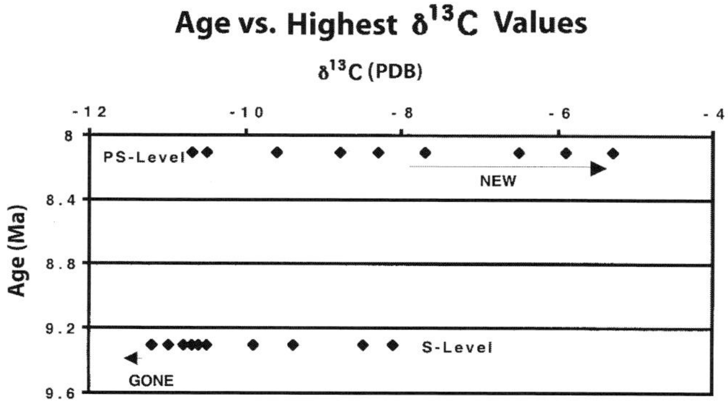

Figure 3.8 Highest $\delta^{13}C$ values per time level. C_4 grasslands appear in the PS-level.

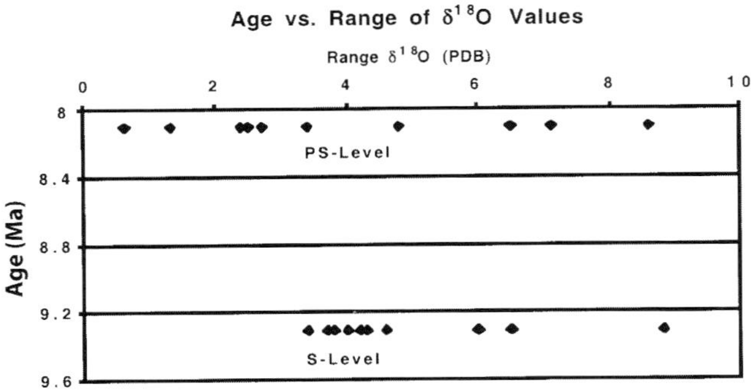

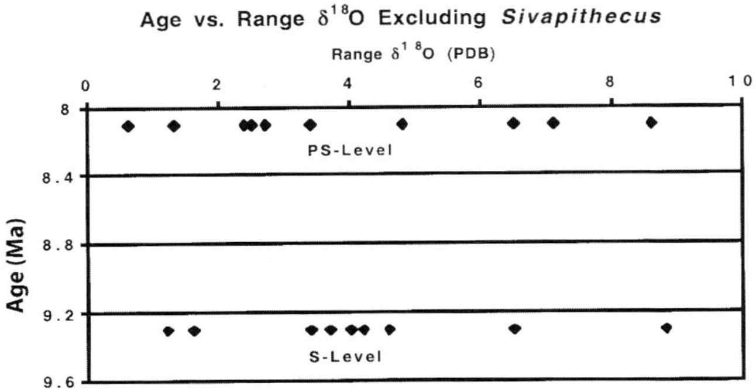

Figure 3.9 Range of δ^{18}O values for each locality in the S-level and the PS-level indicate a similar range of evaporative environments, with localities exhibiting high ranges in both levels. S-level localities do not exhibit small ranges due to the inclusion of *Sivapithecus*, which yields enriched oxygen values in forested environments.

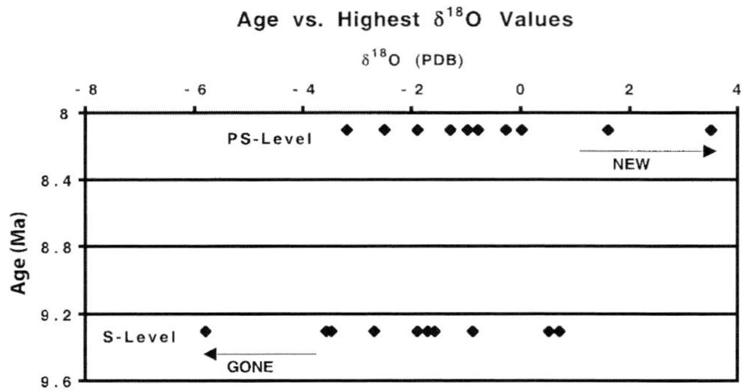

Figure 3.10 Highest δ^{18}O values per time level. Oxygen values become more enriched by the PS-level, suggesting a shift in climate, including a decrease in rainfall.

smaller ranges that some PS-level localities do due to the inclusion of *Sivapithecus*, which exhibits very enriched values in forested environments. When *Sivapithecus* is not included, the spectrum of oxygen ranges per locality is the same for both time levels. However, an enrichment of oxygen values across localities suggests, like the fauna, an oxygen enrichment indicative of a change in climate, including a reduction in annual rainfall (see Figure 3.10). Localities with depleted δ^{18}O values disappear by the PS-level and are replaced by localities with enriched δ^{18}O values.

Conclusions

These isotopic results indicate a shift in vegetation and climate over the course of about 1.1 Ma between the S-level and the PS-level. While a mosaic of habitats is represented in both levels, there appear to have been more closed forest in the S-level and less open habitat compared to the PS-level. While closed forest is still present in the PS-level, much of it has been replaced by open habitat, including some patches of C_4 grass. These C_4 patches of grass were probably not extensive, however, given that at that time no bovids (including a reduncine) had incorporated C_4 graze into their diets. The extinction of S-level taxa with the greatest preferences for closed forest, including *Sivapithecus*, suggests that closed forests, while still present in the PS-level, had become greatly fragmented. Higher ranges of δ^{13}C values for PS-level localities are a further indication of a greater vegetation mosaic by 8.1 Ma.

The enrichment of δ^{18}O values from the S-level to the PS-level indicates a change in rainfall regime. This shift most likely signals a decrease in annual rainfall, possibly in conjunction with other atmospheric or climatic changes. The decrease in annual rainfall suggested by this analysis may have been sufficient to cause the shift in proportions of closed versus open habitat observed, but it was not sufficient to cause major transitions in forest type, and it may not have been enough to allow the sudden competitive advantage of C_4 plants observed. Changes in the rainfall regime will be explored further in Chapter Four.

4 Paleoseasonality Inferred from Equid Teeth and Intra-Tooth Isotopic Variability

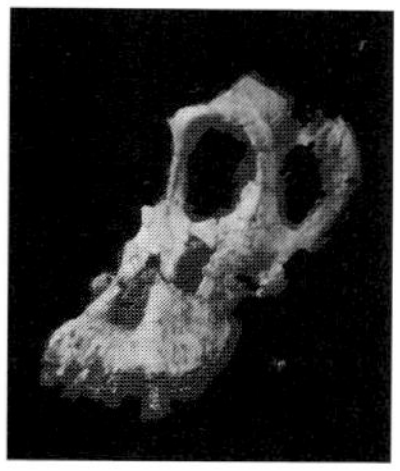

Introduction

Most populations of African apes inhabit evergreen rainforests, with only chimpanzees ranging also into semi-evergreen and deciduous forests in east Africa. In Asia, siamangs and orangutans are limited to evergreen rainforests, with only gibbons ranging into monsoonal rainforests, although even they appear to be restricted to the more diverse and less seasonal forests (Chivers 1984; Kappeler 1984). Modern apes are thus restricted to regions where there is at least some gallery/rainforest and where rainfall usually exceeds 1,500 mm, with dry periods lasting no more than four months.

Boundaries between evergreen, semi-evergreen, and monsoonal moist and dry deciduous forests are largely determined by length of the dry season and total annual rainfall (Walter 1973). Length and severity of the dry season is the primary factor, for forest boundaries are determined by evapotranspiration and availability of soil water (Richards 1996). In India today, evergreen rainforests experience less than four months of a dry season, if any; those with longer dry seasons are also characterized by greater amounts of annual rainfall. Semi-evergreen forests experience three-to-six months of dry season. Moist monsoonal forests have at least four dry months annually but less than seven, and dry monsoon forests are characterized by five-to-eight months of a dry period (Walter 1973).

The purpose of this investigation is to determine whether *Sivapithecus* habitats were characterized by a seasonal rainfall regime, and if so, what kind of forest is associated with this regime in the modern world. Another goal is to determine whether the rainfall regime changed during the later Miocene, for intensification of the Asian monsoon during the Late Miocene could have led to a greater degree of seasonality of precipitation and hence a change in vegetation.

Seasonality of precipitation and/or temperature can be tracked using stable oxygen isotopes. The $\delta^{18}O$ value of precipitation is controlled by a number of factors, including ambient temperature and amount of rainfall. In temperate regions which experience a wide range of temperatures annually, $\delta^{18}O$ values track monthly change in temperature due to differences in evaporative pressure between ^{18}O and ^{16}O, with enriched $\delta^{18}O$ values during summer months. In tropical regions, where rainfall varies but temperature remains relatively constant, $\delta^{18}O$ monthly values track rainfall (Gonfiantini 1985; Njitchoua et al. 1999). ^{18}O precipitates more readily than ^{16}O; therefore, the more it rains, the less ^{18}O is present in atmospheric water, and the more depleted the precipitation becomes (Dansgaard 1964). In monsoonal environments that experience both monthly differences in temperature as well as rainfall, $\delta^{18}O$ values track rainfall, with precipitation greatly depleted during monsoonal months, in spite of high summer temperatures (Araguas-Araguas and Froehlich 1998; Rozanski et al. 1993). For example, on the Tibetan Plateau, summer monsoonal rainfalls are depleted in ^{18}O by more than 6‰ compared to winter rainfall, despite 10 degrees Celsius higher summer temperatures (Araguas-Araguas and Froehlich 1998). Thus, in a monsoonal system, the precipitation amount effect overpowers the temperature effect.

These climate-sensitive changes in meteoric water, or water that participates in atmospheric circulation, then become incorporated into an animal's water source and ultimately into its body water. $\delta^{18}O$ values of tooth enamel reflect the isotopic composition of water ingested by an animal during enamel formation (Longinelli 1984; Luz et al. 1984). Multiple studies have shown that inter- and intra-tooth variation in isotopic composition of carbonate and phosphate oxygen in both modern and fossil taxa represents the intra-annual seasonal oscillations that the animal experienced (Bryant et al. 1994, 1996a,

1996b; Fox and Fisher 2001; Fricke et al. 1998; Fricke and O'Neil 1996; Gadbury et al. 2000; Koch et al. 1989; Sharp and Cerling 1998). Submillimeter samples of tooth enamel taken throughout the length of a hypsodont tooth, such as an equid's which takes a year or more to develop, can therefore provide a record of annual seasonality (Cerling and Sharp 1996; Sharp and Cerling 1996).

Unlike carbon isotopic values, no universal fractionation factors apply between diet and tooth enamel for fauna in general, because fractionation of oxygen isotopes between body and environmental drinking water is dependent on the rates of drinking and respiration (Luz et al. 1984). Furthermore, certain caveats must be considered in inferring precipitation values from tooth enamel. Herbivores, which obtain a significant amount of their water from vegetation, will have $\delta^{18}O$ values that reflect not only the $\delta^{18}O$ value of local precipitation but also values from the leaves that are enriched due to evapotranspiration. Preferential evaporation of ^{16}O can elevate $\delta^{18}O$ values of leaf water by 10‰ or more (Dongmann et al. 1974). Thus while $\delta^{18}O$ values of precipitation are the predominant control over $\delta^{18}O$ values of body water, relative humidity can be a factor in arid and semi-arid environments and with animals that obtain most or all of their body water from vegetation (Ayliffe et al. 1992; Ayliffe and Chivas 1990). These factors can be minimized by analyzing animals that drink daily and have high water turnover rates, such as zebra or horses, who will therefore have a body water composition more sensitive to surface water composition. All Siwalik precipitation regimes are inferred from equid teeth to minimize evapotranspiration effects.

In addition to evapotranspiration effects, $\delta^{18}O$ values of plant water can differ between C_3 plants and C_4 grasses. C_4 plants preferentially undergo evapotranspiration late in the day when it is drier and therefore have higher $\delta^{18}O$ values. In arid settings, this difference can be as high as 10‰ (Sternberg et al. 1984), though in cooler, more humid environments, it is less than 1‰ (Epstein et al. 1977). Kohn et al. (1996) found a positive correlation between increasing C_4 graze content and $\delta^{18}O$ values among grazers and also among browsers and mixed feeders (considered separately from grazers) from east Africa near Lake Turkana. It is unlikely that the Siwaliks were arid during the Late Miocene and the $\delta^{18}O$ difference between C_3 and C_4 plants would therefore have been closer to 1‰ than to 10‰.

Water sources, such as rivers, respond to changes in $\delta^{18}O$ values of precipitation, though their seasonal response may be attenuated with respect to rainfall. River water comprises both surface run-off of precipitation and groundwater. Where surface run-off predominates, river water $\delta^{18}O$ values track rainfall. This signal becomes dampened, however, as groundwater becomes more of a source, for most groundwater bodies are isotopically constant and reflect the average annual isotopic composition of precipitation (Fritz 1981; Gat 1981). Furthermore, in arid zones, evaporation can enrich $\delta^{18}O$ values. A study of rivers in north central Africa, however, suggests that even in arid regions where evaporation is responsible for most seasonal variations, these waters respond strongly to monsoon rains, closely approaching the $\delta^{18}O$ of rainfall during the monsoonal months (Fontes et al. 1970; Fritz 1981). Smaller bodies of water, such as ephemeral drinking pools, experience rapid volume turnover, thereby reducing the effect of evaporation and allowing them to more accurately reflect $\delta^{18}O$ values of precipitation (Koch et al. 1989). Thus, while Siwalik equids that drank from shallow pools are likely to yield $\delta^{18}O$ values reflective of the isotopic composition of rainfall, individuals that drank from larger, buffered bodies of water or from springs will yield an attenuated signal.

Finally, Bryant et al. (1994, 1996a) have suggested the mother's body water and physiology could affect teeth which develop before weaning, with weaning occurring at nine to 15 months of age. In horses and zebras, first molars mineralize before weaning, and second molars mineralize around the age of weaning. All other premolars and molars mineralize after weaning. These authors therefore suggest excluding M1s and M2s from climatic reconstructions, even though they did not find a consistent pattern in $\delta^{18}O$ values from M1 to M3 molars in Oligocene and Miocene horses (Bryant et al. 1996a). Gadbury et al. (2000) also report no differences between molar positions of Early Holocene bison teeth and suggest that there is no significant difference in the $\delta^{18}O$ value of body water obtained from the mother versus water obtained from post-weaning dietary intake.

Reconstructions of Siwalik paleoclimates with respect to rainfall are useful not only for inferring *Sivapithecus*'s habitat preferences but also for determining the timing of inception and strengthening of the Asian monsoon. A monsoon climate is defined by a reversal of wind direction between summer and winter, blowing from cooler oceans to warmer land in the summer and from cooler continents to warmer oceans in the winter. Summer winds bring moisture from the ocean to land, resulting in heavy rainfall, while winters are usually dry. The southeast Asian-southwest Indian monsoon system today envelops the tropical and subtropical continents of Asia, Africa, and Australia. In India, moisture is carried from both the Arabian Sea on southwest winds and the Bay of Bengal on southeast winds during the summer, while winds blow across the dry Asian continent from the northeast in winter (Mooley and Shukla 1987). Pakistan also receives these monsoonal rains, although in the more northern latitudes the monsoon rains arrive later and leave earlier than in more southern latitudes. Pakistan in addition receives supplemental rains during the winter from prevailing westerlies carrying moisture from the Mediterranean, Black, and Caspian seas (Murakami 1987).

The Asian monsoon is driven by three primary forces-the differential heating of the land and ocean, the swirl introduced to winds by rotation of the earth, and by moisture effects which result from the earth's atmospheric temperature and pressure lying very close to the triple point of water (solid, liquid, and gaseous phases coexist in equilibrium) (Webster 1987). The land-ocean temperature gradient is the predominant force, which in turn is driven by the Tibetan Plateau. The plateau generates a high-altitude region of low pressure in the summer as it warms, thus pulling moisture-laden air masses from the ocean, and it subsequently generates a high-pressure region in the winter as it cools, returning moisture from the land back to the ocean.

Climatic modeling suggests that the Tibetan plateau had to be at least half its current height in order to drive a strong monsoonal system (Prell and Kutzbach 1992). The timing of uplift of the plateau remains a point of contention, with estimates ranging from 40 to 3.4 Ma, but a range of evidence suggests that while uplift started around 50 Ma, a significant increase in altitude began around 10–8 Ma (Zhisheng et al. 2001).

Foraminiferal records indicate the beginning of strong summer Arabian upwellings around 9.4 Ma, with continued strengthening until around 8.2 Ma, suggesting the beginning or at least major intensification of the Asian monsoon (Kroon et al. 1991, corrected to Cande and Kent 1995). At 9.15 Ma in the Siwaliks of Pakistan, paleosol data indicate a shift in $\delta^{18}O$ values, with samples becoming more enriched over time, especially after 8.0 Ma, suggesting a decrease in rainfall and/or increasing seasonality of rainfall (Quade and Cerling 1995; Quade et al. 1989). This drying in Pakistan is predicted in climatic modeling as mountains elevate and precipitation increases in east Asia (Prell and Kutzbach 1992).

Evidence from the Nepal Siwaliks also indicates decreasing rainfall in the Late Miocene. Dettman et al. (2001) tracked $\delta^{18}O$ values in fossil bivalves from the Siwalik sediments in Nepal from the past 10.7 million years and found evidence for wet-dry seasons in all shells, suggesting that the monsoon system was already in effect as early as 10.7 Ma. Furthermore, since 9.5 Ma, the magnitude of seasonal variability in $\delta^{18}O$ values remained essentially unchanged. The $\delta^{18}O$ values before 7.5 Ma were more depleted than younger specimens, suggesting a decrease in rainfall beginning at 7.5 Ma. The decrease in rainfall in Pakistan, as inferred from the oxygen isotope shift observed in paleosols, therefore precedes the shift observed in Nepal by 0.5 Ma.

Materials and Methods

A total of 30 *Hipparionini* teeth spanning 10.3 to 6.3 Ma were sampled from the Siwalik record for carbon and oxygen isotopes. Fifteen of these teeth are represented by intra-tooth profiles, with each profile consisting of 13 to 23 samples spanning 33 to 53 mm from occlusal surface to root. Samples were removed from teeth using a Dremel high-speed rotary tool with a tungsten-carbide or diamond-impregnated bit. Tooth profiles were analyzed using the automated carbonate microsampler method described in Chapter Three. The remaining teeth were analyzed cryogenically, as also described in Chapter Three.

In two cases, the same tooth was sampled using both methods, and in two other cases, two individual animals are represented by two different teeth from the same jaw. All cases yielded similar carbon results, and all but one (Y50803) yielded the same oxygen results, as summarized in Chapter Three. Profiles include third and fourth premolars and first, second, and third molars.

For comparison, intra-jaw profiles were sampled in seven bovid specimens. Four *Tragoceridus* spp. and three *Dorcadoxa porrecticornis* jaws were sampled on both the second and third molar, one sample per tooth. Multiple samples were not taken from the same tooth because these bovid teeth are not hypsodont and therefore develop much more quickly compared to hypsodont equid teeth. The second and third molars were sampled for two reasons-first, to avoid potential weaning effects that might be associated with M1s, and second, because M2s develop earlier than any of the remaining premolars or molars, thus allowing for greatest time difference between intra-jaw samples (Hillson 1986). Additionally, Miocene equid teeth were sampled from the Chorrora formation in Ethiopia, dated to 10.8 Ma, and the

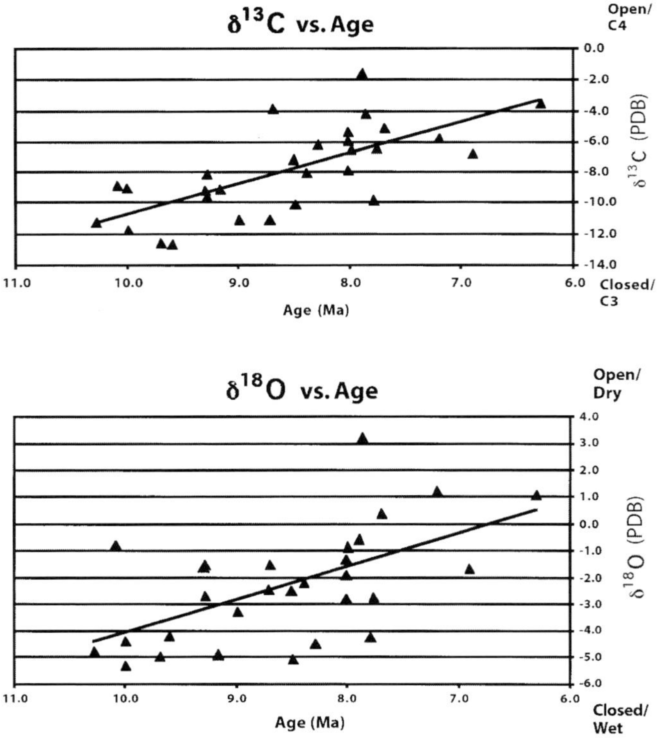

Figure 4.1 $\delta^{13}C$ and $\delta^{18}O$ values become more enriched over time, indicating a transition from closed/ relatively wet habitat to open/dry habitat for horses. Carbon: Adjusted R^2=.48; p<.0001. Oxygen: Adjusted R^2=.32; p=.0009.

Dinotheriesandes in Germany, dated to 10.5 Ma. Each of the two regions is represented by four teeth, with two of those teeth sampled three times along the tooth from crown to root. All comparative specimens were analyzed using the automated carbonate microsampler method.

Isotopic ratios are presented in the per mil (‰) notation $\delta^{13}C$ (or $\delta^{18}O$) = (R_{sample}/R_{PDB}−1) x 1000 where R_{sample} and R_{PDB} are the ratios $^{13}C/^{12}C$ (or $^{18}O/^{16}O$) in the sample and standard respectively, and the isotope reference standard is PDB (Pee Dee Belemnite). Oxygen values are reported assuming the acid-calcite fractionation factor for calcite. All samples were analyzed at the University of Utah.

Results and Discussion

Compilation of Equid Teeth Sampled

When all horse teeth are compiled, with average $\delta^{13}C$ and $\delta^{18}O$ values reported for tooth profiles, and individuals sampled by both methods represented only by the profile averages to avoid repetition, there are 29 *Hipparion* teeth which span 4 Ma (10.3–6.3 Ma) (see Table 4.1). Among these teeth, both carbon and oxygen values correlate negatively with age (Spearman Rank Correlation: carbon t^* = −.687; p = .000; oxygen t^* = −.509; p = .005), indicating a shift to more open and drier habitats over time (see Figure 4.1). This conclusion is further supported by a positive correlation between carbon and oxygen values (t^* = .719; p = .000), indicating open habitats are also drier, while closed habitats are wetter or are undergoing less evaporative effects (see Figure 4.2).

Different teeth suggest a range of habitats exploited by Siwalik equids. Two teeth dated to 9.7 and 9.6 Ma have carbon values less than −12‰, suggesting an affinity for closed habitat. The first clear evidence of C_4 graze appears at 8.7 Ma, with a range of −4.4 to −3.1‰ in this profiled tooth (Y3699). While these values indicate some C_4 graze in the diet throughout the entire tooth, they also reflect a strong C_3 component as well. Teeth dated to 10.1 and 9.3 Ma also show enriched $\delta^{13}C$ values (−7.5 and −7.6‰ respectively) which could reflect a small component of C_4 graze or alternatively, a diet of C_3 plants in open habitats. Teeth subsequent to the first evidence of C_4 graze at 8.7 Ma, or those sampled at 8.5, 8.4, and 8.3 Ma, indicate periods of the year in which

Table 4.1

Compilation of equid teeth sampled

Specimen ID	Locality	Age (Ma)	$\delta^{18}O$	$\delta^{13}C$
Y27729	Y729	10.3	-4.8	-11.3
Y9488	Y258	10.1	-0.8	-8.8
MM120	Y311	10.0	-4.4	-9.0
Y11881	Y311	10.0	-5.3	-11.7
Y46669	Y875	9.7	-4.9	-12.5
Y49603	Y893	9.6	-4.2	-12.6
AVERAGE			**-4.0**	**-11.0**
ST DEV			**1.7**	**1.7**
Y19810	Y312	9.3	-1.6	-9.2
Y19813	Y312	9.3	-1.5	-9.5
Y47080	Y312	9.3	-2.6	-8.1
Y46248	Y418	9.2	-4.9	-9.1
Y5097	Y210	9.0	-3.2	-11.1
Y49666	RK03	8.7	-2.4	-11.1
Y3699	Y158	8.7	-1.5	-3.8
Y18378	Y608	8.5	-2.5	-7.1
Y15610	Y467	8.5	-5.1	-10.0
Y15403	Y443	8.4	-2.2	-7.9
Y3050	Y113	8.3	-4.5	-6.1
AVERAGE			**-2.9**	**-8.5**
ST DEV			**1.3**	**2.2**

Specimen ID	Locality	Age (Ma)	$\delta^{18}O$	$\delta^{13}C$
Y49674	Y542	8.0	-1.3	-5.3
Y18418	Y606	8.0	-1.9	-7.7
Y17859	Y599	8.0	-2.7	-5.9
Y49548	Y890	8.0	-0.8	-6.5
Y579	Y029	7.9	-0.6	-1.4
Y49518	Y886	7.9	3.2	-4.1
Y50016	Y917	7.8	-4.3	-9.8
Y50870	Y948	7.8	-2.7	-6.4
Y15293	Y435	7.7	0.4	-5.0
Y50130	Y932	7.2	1.2	-5.7
Y50803	Y925	6.9	-1.6	-6.7
Y49784	Y912	6.3	1.1	-3.4
AVERAGE			**-0.8**	**-5.7**
ST DEV			**2.1**	**2.1**

Compilation of 29 *Hipparion* teeth, with tooth profiles represented by average isotopic values, and individuals sampled by both the profile method and the single sample method represented only by the profile averages to avoid repetition. $\delta^{13}C$ and $\delta^{18}O$ transitions cluster with respect to three age groups—10.3-9.6 Ma, 9.3–8.3 Ma, and 8.0–6.3 Ma. Both carbon and oxygen transitions indicate a change in horse habitat from wooded to open habitat with C_4 graze, and from wetter/ more humid climate to drier.

no C_4 component is present, suggesting C_4 patches were small or greatly dispersed. It is not until 8.0 Ma that all horse profiles yield evidence of a C_4 component throughout the entire tooth, suggesting permanent patches of C_4 grass. Still, a 7.8 Ma tooth (Y50016) sampled once yielded depleted carbon (–9.8‰) and oxygen (–4.3‰) values, suggesting some horses continued to feed only on C_3 plants at least part of the year in somewhat less open and wetter or less evaporative habitat.

Like carbon values, oxygen values are more depleted in older specimens than in younger. In six teeth dating to 9.6 Ma or older, all but one have $\delta^{18}O$ values less than –4‰ (see Table 4.1). The exception (Y9488 from locality Y258) is 10.1 Ma, and its enriched $\delta^{13}C$ values

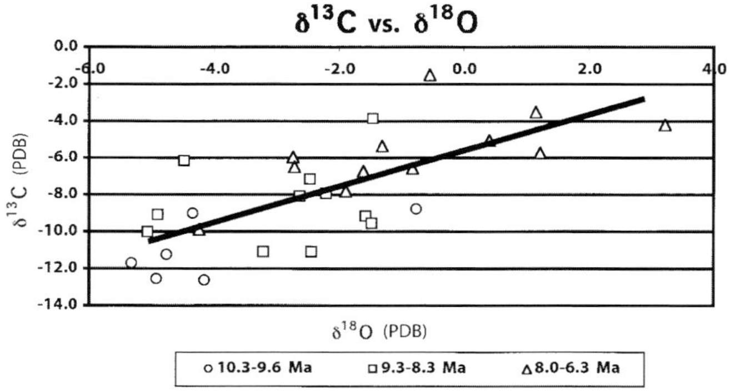

Figure 4.2 $\delta^{13}C$ and $\delta^{18}O$ values show a positive correlation, indicating more closed habitats are wetter or less evaporative, while more open habitats are drier or more evaporative. Linear regression: adjusted R^2=.48; p<.0001.

throughout the length of the tooth (–9.8 to –7.5‰) indicate feeding in open habitat where evaporative effects could result in enriched $\delta^{18}O$ values, both from evaporated water bodies and open-habitat vegetation. These values are consistent with lithological evidence that this locality was unusually open, resembling localities in the Dhok Pathan Formation interfan river system even though it came from the Nagrii Formation large emergent river system. Thus isotopic evidence from locality Y258 suggests that vegetation responded to specific local conditions, which would have resulted in local vegetation mosaics as well as mosaics spanning several kilometers.

The first enriched $\delta^{18}O$ values appear consistently at 9.3 Ma, although depleted values are also present consistently until 8.3 Ma. Eleven teeth fall between 9.3 and 8.3 Ma, with four teeth yielding $\delta^{18}O$ values below –3‰. From 8.0 Ma and later, of 12 teeth, all but one tooth has become enriched (greater than –3‰) with

Table 4.2

Mann-Whitney comparisons of equid carbon values per age group

	10.3-9.6 Ma	9.3-8.3 Ma
10.3-9.6 Ma		
9.3-8.3 Ma	Z=-2.11; p=.035	
8.0-6.3 Ma	Z=-3.18; p=.002	Z=-2.71; p=.007

Mann-Whitney pair-wise comparisons of $\delta^{13}C$ values for three age groupings of *Hipparion* teeth indicate significant shifts from wooded to more open habitat and some C_4 graze, followed by another transition to more open habitat and more C_4 graze.

Table 4.3

Mann-Whitney comparisons of equid oxygen values per age group

	10.3-9.6 Ma	9.3-8.3 Ma
10.3-9.6 Ma		
9.3-8.3 Ma	Z=-1.26; p=ns	
8.0-6.3 Ma	Z=-2.37; p=.008	Z=-2.37; p=.018

Mann-Whitney pair-wise comparisons of $\delta^{18}O$ values for three time intervals of *Hipparion* teeth indicate an initial transition from more closed/wet/less evaporative habitat to a mixture of closed and open or wet and dry habitats, followed by a second transition to open/dry habitat. These vegetation shifts were most likely accompanied by a decrease in annual rainfall over time.

consistently highly enriched values, but again the tooth sampled once at 7.8 Ma which was mentioned earlier does have a depleted oxygen value.

These results suggest that while some open habitat and some more wooded habitats were present throughout the four million years sampled, a transition to a greater proportion of open, drier habitats occurred first between 9.6 and 9.3 Ma, with a second transition to even more open and drier habitats between 8.3 and 8.0 Ma (see Table 4.1). This vegetation shift was most likely accompanied by a decrease in rainfall as well, and variability in oxygen values between 9.3 and 8.3 Ma may represent variability in climate. When the samples are divided into three age groups—10.3–9.6 Ma, 9.3–8.3 Ma, and 8.0–6.3 Ma-Kruskal-Wallis 1-Way Anovas corrected for ties indicate significant differences between the groups for carbon (X^2 = 14.6; p = .0007) and oxygen (X^2 = 9.88; p = .007). For carbon values, all pair-wise Mann-Whitney comparisons between the three age groups yield significant differences, suggesting first a transition from wooded to more open habitats for horses, followed by a transition to open habitat including patches of C_4 grass (see Table 4.2). For oxygen values, the first transition between oldest and middle ages is not significant, indicating that while open habitat values become more predominant, horses are still taking advantage of more wooded habitat as well. Additionally, amount of annual rainfall may have been variable. The second transition, however, from the middle age group to the youngest, yielded a significant increase in $\delta^{18}O$ values (Z = –2.37; p = .018), indicating a shift away from wooded habitat and a permanent decrease in rainfall (see Table 4.3).

Finally, when the 29 teeth are divided into those present before the extinction of *Sivapithecus* (10.3–8.4 Ma, n = 16) and those present afterwards (8.3–6.3 Ma, n = 13), there is a significant enrichment in carbon values with time, from an average $\delta^{13}C$ value of –9.5‰, indicating no graze on average, to an average of –5.7‰, indicating some graze on average (Z = –3.68; p = .0002). Likewise, there is a significant enrichment of $\delta^{18}O$ values from –3.2‰ on average to –1.1‰ on average (Z = –2.55; p = .011). This 2.1‰ increase is consistent with the 2.3‰ increase seen in the all-inclusive taxa comparison between the S-level and the PS-level discussed in Chapter Three.

This consistency suggests that mammalian teeth across a range of taxa are detecting a shift in the rainfall regime as well as a shift in the vegetation mosaic.

Intra-tooth Sampling Profiles

Fifteen *Hipparionini* teeth ranging in age from 10.1 to 6.3 Ma were sampled throughout the length of the tooth (see Appendix 2 and Figures 4.3–4.17). These teeth included P3s, P4s, and all molars. Kruskal-Wallis 1-Way Anovas indicated no significant differences in high, low, mean, or range of carbon or oxygen values based on tooth position, suggesting there were no weaning effects on $\delta^{18}O$ values. In fact, M1s and M3s yielded overlapping oxygen curves in comparisons of a 10.0 Ma M3 and an 8.5 Ma M1 and again between a 7.2 Ma M3 and a 6.3 Ma M1 (see Figure 4.18).

As in the 29 horse teeth sample, in the 15 profiles both carbon and oxygen become enriched over geological time (Spearman Rank correlation: carbon: $t^* = -.73$; $p = .002$; oxygen: $t^* = -.61$; $p = .015$) (see Figures 4.19 and 4.20). Furthermore, average $\delta^{13}C$ and $\delta^{18}O$ values for each tooth show a positive correlation ($t^* = .78$; $p = .001$). This correlation suggests that over time, amount of annual rainfall decreased, changes in contributions from different precipitation sources may have occurred, and habitats became more open. However, within individual profiles, there is not a consistent relationship between carbon and oxygen values. Trends within teeth include increasing $\delta^{13}C$ with increasing $\delta^{18}O$, decreasing $\delta^{13}C$ with increasing $\delta^{18}O$, or no pattern at all. The fact that carbon and oxygen did not correlate or show consistent trends within teeth suggests that the $\delta^{18}O$ curves are not indicative of movement from closed to open habitat within the time span of tooth development, for if that were the case, carbon and oxygen should show a positive correlation. Rather, the oxygen curves are indicative of annual changes in precipitation.

Intra-tooth variation of carbon values suggests that some equids exploited a greater range of habitats than others. Intra-tooth ranges in $\delta^{13}C$ values fell between 0.3 and 2.6‰ (see Table 4.4; Figure 4.19). Those teeth showing the least variability were: 1) the three teeth with the most depleted $\delta^{13}C$ values (at 10.0, 9.7, and 9.0 Ma), indicating an affinity for more closed habitat and diets of pure C_3 with little ranging into more open

Table 4.4

Mean carbon and oxygen values per equid tooth

Age	Mean $\delta^{13}C$	Range $\delta^{13}C$	Mean $\delta^{18}O$	Range $\delta^{18}O$
10.1	-8.8	2.3	-0.8	2.5
10	-11.7	0.9	-5.3	3.0
9.7	-12.5	0.6	-4.9	2.1
9.3	-9.2	2.3	-1.6	2.4
9	-11.1	1.1	-3.2	3.4
8.7	-3.8	1.3	-1.5	3.4
8.5	-10.0	1.4	-5.1	2.2
8.4	-7.9	2.0	-2.2	2.7
8.3	-6.1	3.6	-4.5	2.2
8	-6.5	2.7	-0.8	3.9
7.9	-1.4	1.0	-0.6	2.0
7.7	-5.0	2.4	0.4	3.3
7.2	-5.7	1.6	1.2	2.1
6.9	-6.7	2.1	-1.6	3.1
6.3	-3.4	2.9	1.1	2.9
AVERAGE		**1.9**		**2.7**
STDEV		**0.83**		**0.57**

Mean $\delta^{13}C$ (PDB) and $\delta^{18}O$ (PDB) values per profiled tooth, with ranges of values per tooth. Teeth showing the least variability in carbon values, and hence more restricted use of the vegetation mosaic, are those with very depleted carbon and oxygen values, indicative of forest, and the tooth most enriched in carbon, indicative of a predominantly C_4 diet.

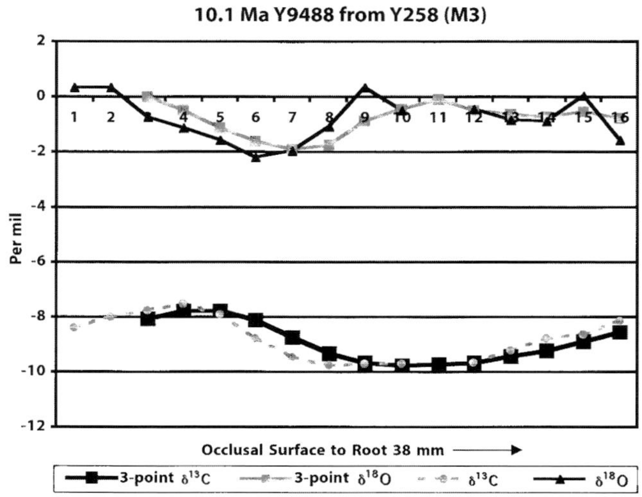

Figure 4.3 Carbon (bottom curve) and oxygen (top curve) intra-tooth sampling profiles of a 10.1 Ma *Hipparion* tooth. Three-point averaged curves in bold, original data in dashed-lines.

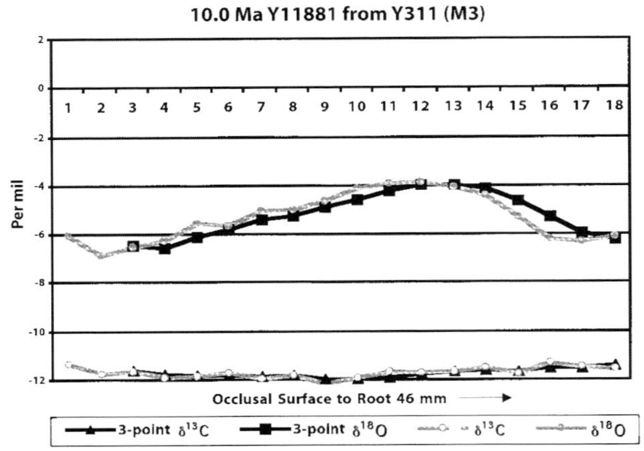

Figure 4.4 Carbon (bottom curve) and oxygen (top curve) intra-tooth sampling profiles of a 10.0 Ma *Hipparion* tooth. Three-point averaged curves in bold, original data in dashed-lines.

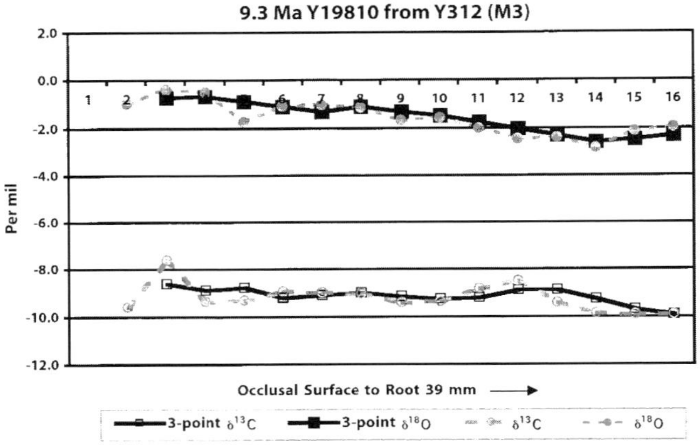

Figure 4.6 Carbon (bottom curve) and oxygen (top curve) intra-tooth sampling profiles of a 9.3 Ma *Hipparion* tooth. Three-point averaged curves in bold, original data in dashed-lines.

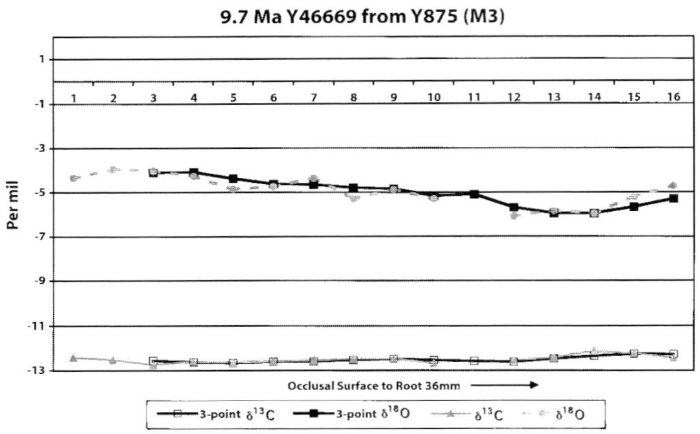

Figure 4.5 Carbon (bottom curve) and oxygen (top curve) intra-tooth sampling profiles of a 9.7 Ma *Hipparion* tooth. Three-point averaged curves in bold, original data in dashed-lines.

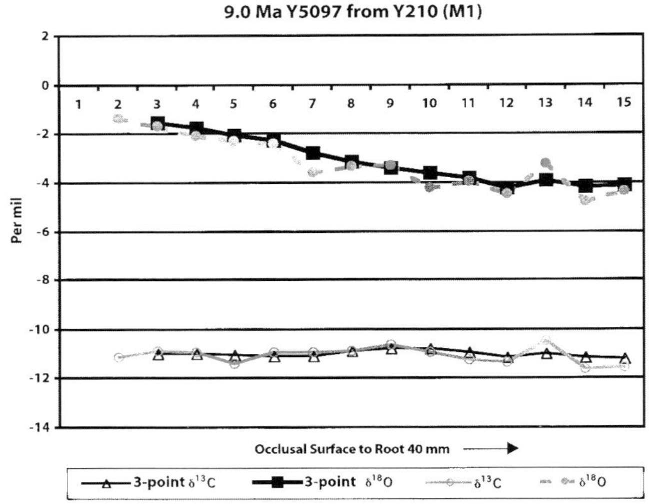

Figure 4.7. Carbon (bottom curve) and oxygen (top curve) intra-tooth sampling profiles of a 9.0 Ma *Hipparion* tooth. Three-point averaged curves in bold, original data in dashed-lines.

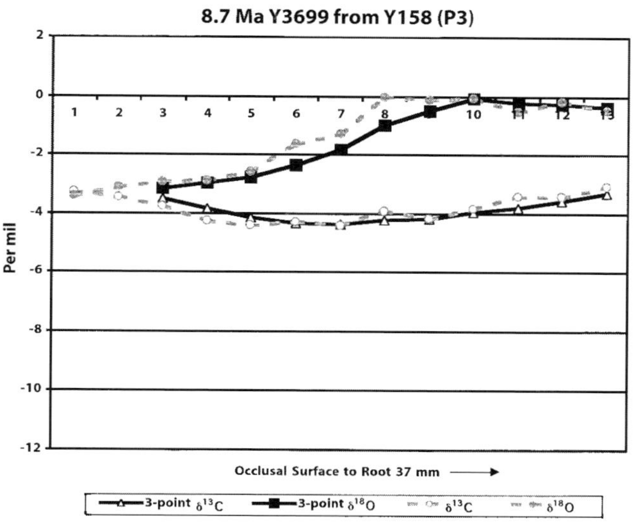

Figure 4.8 Carbon (bottom curve) and oxygen (top curve) intra-tooth sampling profiles of a 8.7 Ma *Hipparion* tooth. Three-point averaged curves in bold, original data in dashed-lines.

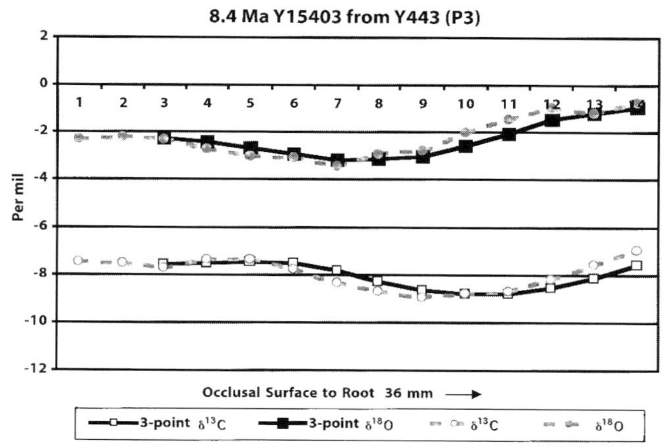

Figure 4.10 Carbon (bottom curve) and oxygen (top curve) intra-tooth sampling profiles of a 8.4 Ma *Hipparion* tooth. Three-point averaged curves in bold, original data in dashed-lines.

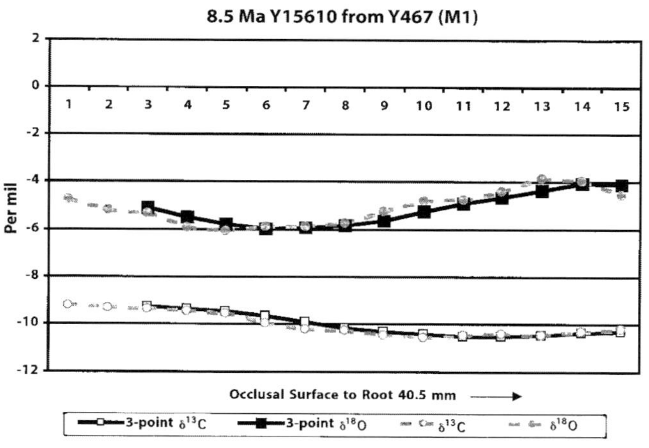

Figure 4.9 Carbon (bottom curve) and oxygen (top curve) intra-tooth sampling profiles of a 8.5 Ma *Hipparion* tooth. Three-point averaged curves in bold, original data in dashed-lines.

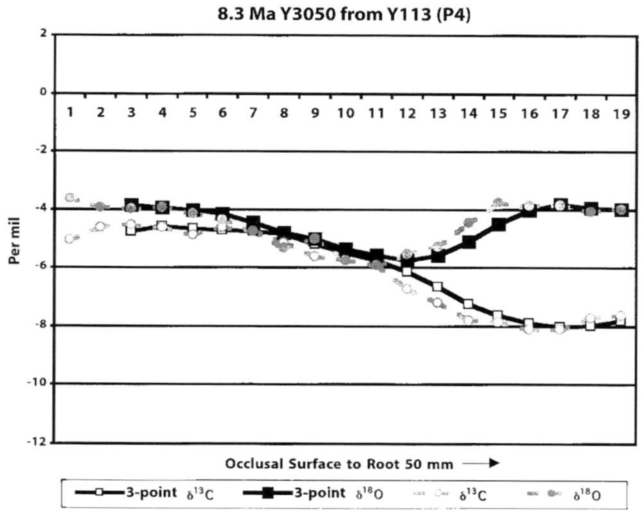

Figure 4.11 Carbon (bottom curve) and oxygen (top curve) intra-tooth sampling profiles of a 8.3 Ma *Hipparion* tooth. Three-point averaged curves in bold, original data in dashed-lines.

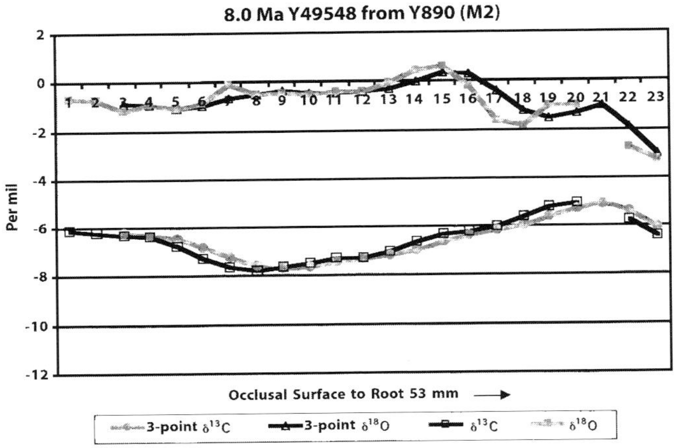

Figure 4.12 Carbon (bottom curve) and oxygen (top curve) intra-tooth sampling profiles of a 8.0 Ma *Hipparion* tooth. Three-point averaged curves in bold, original data in dashed-lines.

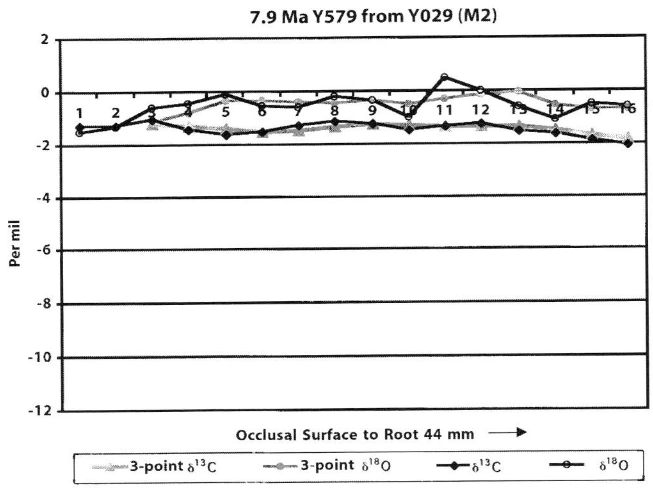

Figure 4.13 Carbon (bottom curve) and oxygen (top curve) intra-tooth sampling profiles of a 7.9 Ma *Hipparion* tooth. Three-point averaged curves in bold, original data in dashed-lines.

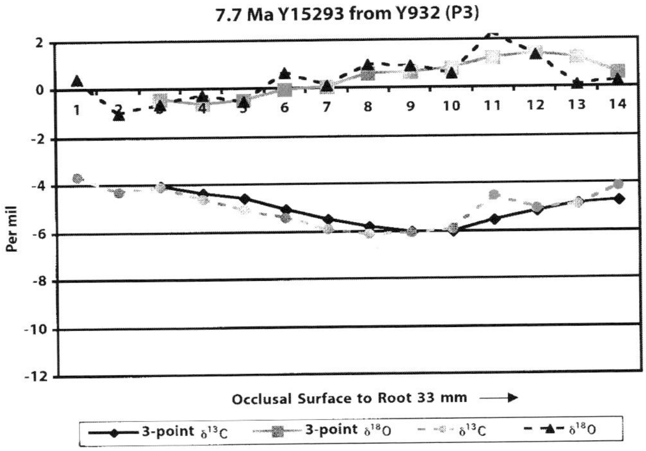

Figure 4.14 Carbon (bottom curve) and oxygen (top curve) intra-tooth sampling profiles of a 7.7 Ma *Hipparion* tooth. Three-point averaged curves in bold, original data in dashed-lines.

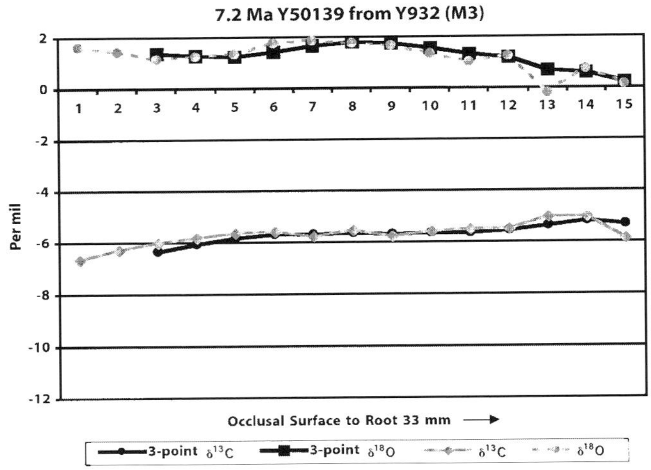

Figure 4.15 Carbon (bottom curve) and oxygen (top curve) intra-tooth sampling profiles of a 7.2 Ma *Hipparion* tooth. Three-point averaged curves in bold, original data in dashed-lines.

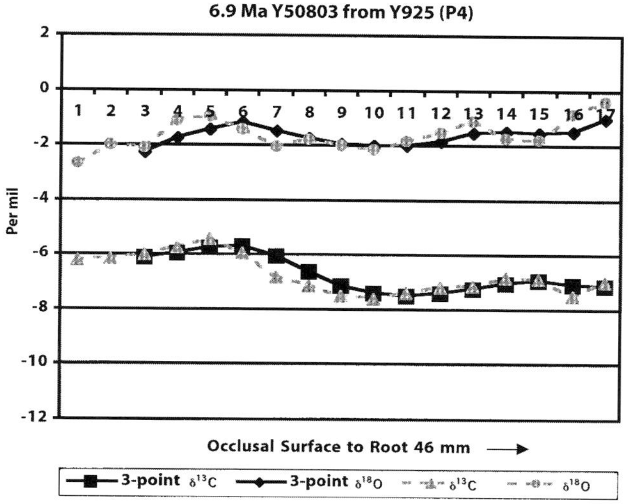

Figure 4.16 Carbon (bottom curve) and oxygen (top curve) intra-tooth sampling profiles of a 6.9 Ma *Hipparion* tooth. Three-point averaged curves in bold, original data in dashed-lines.

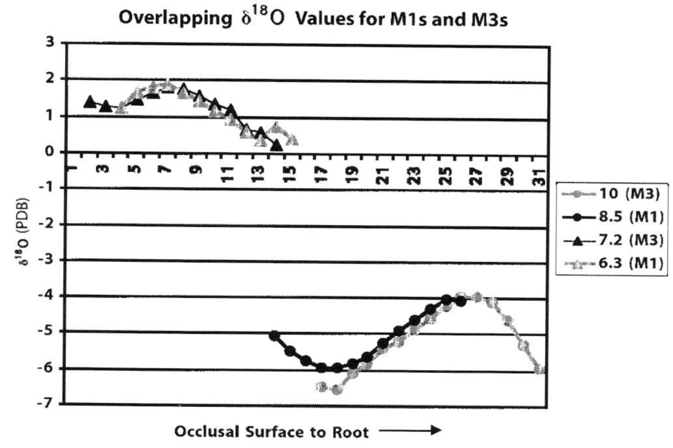

Figure 4.18 M1s and M3s yield similarly shaped $\delta^{18}O$ curves as well as similar values, indicating no weaning effect. Profiles are represented by three-point running averages.

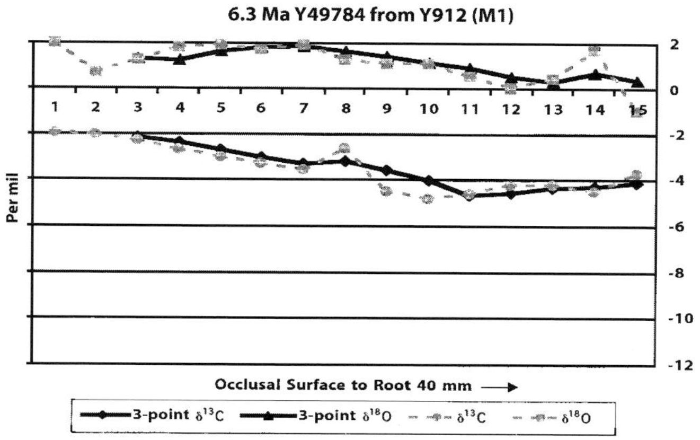

Figure 4.17 Carbon (bottom curve) and oxygen (top curve) intra-tooth sampling profiles of a 6.3 Ma *Hipparion* tooth. Three-point averaged curves in bold, original data in dashed-lines.

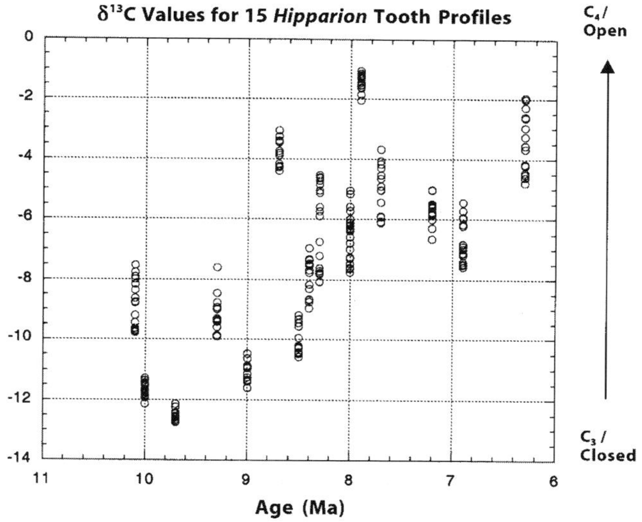

Figure 4.19 $\delta^{13}C$ (PDB) values for horse profiles indicate enrichment, and hence more open habitat and C_4 graze, over time. Each column of circles represents all samples from one tooth. The most depleted teeth, or those in relatively closed habitat, have the smallest ranges in carbon values. The first clear evidence for incorporation of C_4 graze into horse diets (greater than –7.5‰) appears at 8.7 Ma, though all teeth have a C_3 component. Note the 10.1 Ma tooth is remarkably enriched compared those at 10.0 and 9.7 Ma. This tooth is from locality Y258, that lithologically resembles younger, more open habitats.

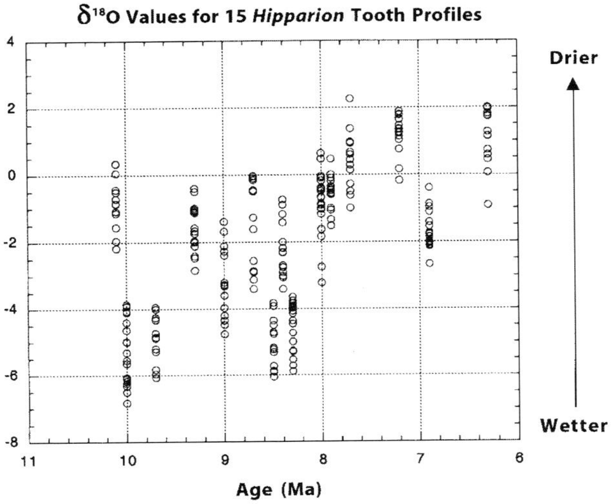

Figure 4.20 $\delta^{18}O$ (PDB) values for horse tooth profiles show a trend towards enrichment over time, suggesting a decrease in amount of annual rainfall and increase in open habitat. Each column of circles represents all samples from one tooth. Range of oxygen values is consistently around 2.7‰ per tooth. Note the 10.1 Ma tooth is remarkably enriched compared to teeth at 10.0 and 9.7 Ma. This tooth is from locality Y258, which lithologically resembles younger, more open habitats as well.

habitats, and; 2) the tooth with the most enriched values (7.9 Ma), which indicate the heaviest component of C_4 graze in the diet. Those teeth with the greatest ranges in carbon indicate mixed diets of C_3 and C_4. Using –7.5‰ as the most depleted $\delta^{13}C$ values, which could indicate a portion of C_4 graze in the diet, the first clear evidence of C_4 graze occurs at 8.7 Ma, as mentioned earlier. This tooth indicates a mix of C_3 and C_4 throughout the length of the tooth. Younger teeth at 8.5 and 8.4 Ma yield little to no evidence of C_4 graze. At 8.3 and 8.0 Ma, evidence of C_4 graze reappears, but for each tooth, there are periods of the year in which C_4 may not have been a component of the diet. All younger teeth, beginning with a 7.9 Ma specimen, indicate a diet of both C_3 and C_4 throughout the length of the tooth. The most enriched tooth, at 7.9 Ma, yields average $\delta^{13}C$ values of –1.4‰, with a high of –1.1‰, still indicating a small component of C_3. Modern hyper-grazers (a diet of 100 percent C_4) yield C values of +1 to +4‰ that become even more enriched when 1.5‰ is added in order to compensate for the shift in isotopic ratios of atmospheric CO_2 in the past 150 years due to the burning of fossil fuels. Thus no Siwalik equids have become 100 percent grazers by 6.3 Ma.

Intra-tooth variability in oxygen values follows a trend among consecutive samples most likely representing seasonal fluctuations in precipitation. One tooth (8.3 Ma, Y3050) has an oxygen curve which incorporates a distinguishable, complete seasonal cycle (see Figure 4.11), beginning with enriched values representing the dry season, followed by depleted values indicative of the wet season, followed by a return to enriched values and the dry season. This annual regime was captured over a tooth length of 50 mm. One other tooth (8.0 Ma, Y49548) showing very little to no wear, was sampled over 53 mm, and it yields a complete cycle in its carbon results, but its oxygen curve is more difficult to interpret (see Figure 4.12). Teeth sampled over 45 mm captured most of the cycle, but all other teeth, sampled over shorter lengths, captured only a portion of the cycle, either the dry season before the onset of heavy rains, or the rainy season and subsequent drying (see Figures 4.21 and 4.22).

The 45–50 mm/year rate of enamel formation suggested by Siwalik teeth is greater than the 35–40 mm/year rates described in two other horse and bison tooth studies (Fricke and O'Neil 1996; Sharp and Cerling 1998). In those studies, young adult or unerupted teeth were analyzed, whereas this Siwalik study used worn teeth (except for the 8.0 Ma tooth). This difference in time representation suggests that either Siwalik equids differed from modern equids and bison in rate of enamel formation or that amount of time sampled per length of tooth is dependent upon which portion of the tooth is analyzed. Striae of Retzius, or enamel bands, are laid down at a constant rate, but because the occlusal surface of tooth crowns develops at a faster rate than the remaining tooth, striae are laid down at steeper angles in the crown, or more parallel to the enamel surface. Angle of striae then decreases from crown to root and becomes more perpendicular to the enamel surface (Beynon et al. 1998; Fitzgerald 1998; Hillson 1986; Shellis 1998). This means that surficial sampling of enamel, with each sample covering several millimeters of enamel drilled down almost to the dentine layer, could capture more striae, and hence more time, at the crown. Worn teeth, while requiring more tooth surface to capture a full seasonal cycle, have an advantage in that the original occlusal surface with more angled striae has been worn away so that each sample incorporates less striae, and hence less time-averaging occurs per isotopic sample.

The oxygen isotopic curves from the 11 teeth spanning different seasons (those to the left of the trough, and those that incorporate the trough and to the right) can be combined to form complete seasonal cycles (see Figure 4.23). Justification for this combination is threefold. First, when combined, the curves from the 9.7 Ma tooth and the 8.5 Ma tooth overlap by seven samples along the trough (peak of the wet season), and these consecutive samples yield the same $\delta^{18}O$ values (see Figure 4.24). Tooth profiles, which also overlap in values and closely track the same trajectory include the 10.0 and 8.5 Ma teeth, the 8.7 and 8.4 Ma teeth, and the 7.2 and 6.3 Ma teeth (see Figures 4.21 and 4.22). Furthermore, the resulting combined curves are very similar to the curve from the one tooth (8.3 Ma) that yielded the full seasonal cycle (see Figures 4.11 and 4.23). The 8.3 Ma tooth curve is slightly shorter in length (about 15 samples as opposed to around 19), but this it to be expected given the time-averaging which would result from a single tooth due to sampling multiple striae of Retzius, as mentioned earlier. The fact that teeth from different individuals spanning millions of years follow the same trajectories sample by sample and in many cases yield the same $\delta^{18}O$ values sample by sample indicates that despite caveats in interpreting oxygen isotopes, and potential for variability between teeth due to horse behavior or tooth histology, intra-tooth sampling of oxygen isotopes in hypsondont teeth can yield reconstructions of paleoprecipitation regimes.

Aside from the 11 teeth, which yielded the same $\delta^{18}O$ curve or part of it, four teeth did not fit this pattern (See Figure 4.25). Differences could have arisen at any or all of three of the following levels: 1) sampling the tooth could have resulted in greater time-averaging; 2) individual horse behavior may have differed, including more migratory patterns, buffered water sources, or a greater contribution of vegetation to body water; and 3) there could have been differences in amount of rainfall or length of the wet or dry seasons in the particular year represented by the specimen. The 10.1 Ma tooth is the oldest tooth sampled and shows very enriched carbon and oxygen isotope values compared to other teeth of similar age (See Figures 4.19 and 4.20), suggesting unusually open habitat for that time level. This tooth has an amplitude similar to other teeth but it appears to demonstrate the full annual

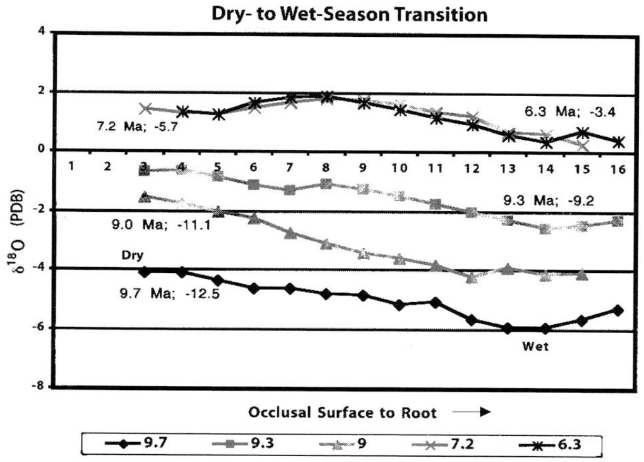

Figure 4.21 Five teeth yielded similarly shaped $\delta^{18}O$ curves which represent the left side of the full annual curve, or a shift from the driest season to the wettest. These teeth range in age from 9.7 Ma to 6.3 Ma, as noted by each curve. Average $\delta^{13}C$ values per curve are also listed beside the age. Average $\delta^{18}O$ values from these curves span a range of 6‰, with a general shift from depleted oldest values to enriched youngest, but values better correlate with $\delta^{13}C$ values. These results suggest a non-monotonic shift to more open habitat over time, which probably accompanied a shift to less annual rainfall. Profiles are represented by three-point running averages.

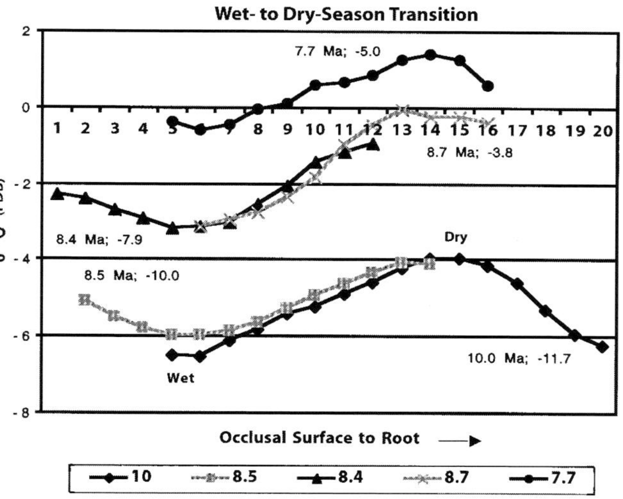

Figure 4.22 Five teeth yielded similarly shaped $\delta^{18}O$ curves which represent the trough and right side of the full annual curve, or a shift from the wettest season to the driest. These teeth range in age from 10.0 Ma to 7.7 Ma, as noted by each curve. Average $\delta^{13}C$ values per curve are also listed beside the age. Average $\delta^{18}O$ values from these curves span a range of 5.7‰, with the oldest value being the most depleted and the youngest, the most enriched. $\delta^{18}O$ values show a positive correlation with $\delta^{13}C$ values. These results suggest a non-monotonic shift to more open habitat over time, which probably accompanied a shift to less annual rainfall. Profiles are represented by three-point running averages.

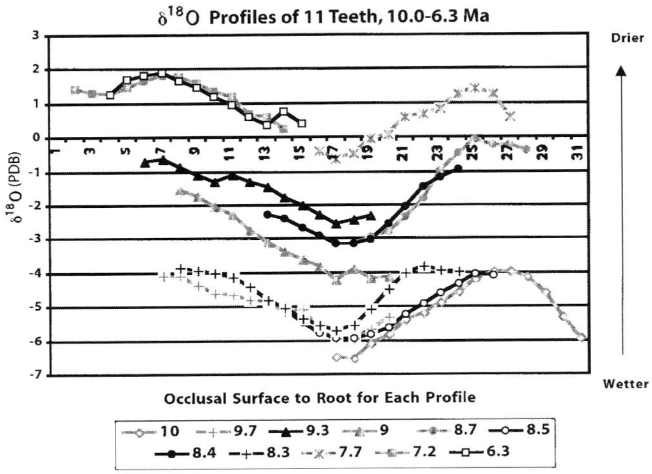

Figure 4.23 When $\delta^{18}O$ profiles are combined for the 10 teeth which made up half of the annual seasonal cycle and the one tooth (8.3 Ma) which incorporated the full annual cycle, a full annual cycle results which becomes enriched over time, suggesting a decrease in rainfall, while maintaining the same shape, implying the same seasonal regime. Profiles are represented by three-point running averages. For any given curve, crests represent dry seasons and troughs represent rainy seasons.

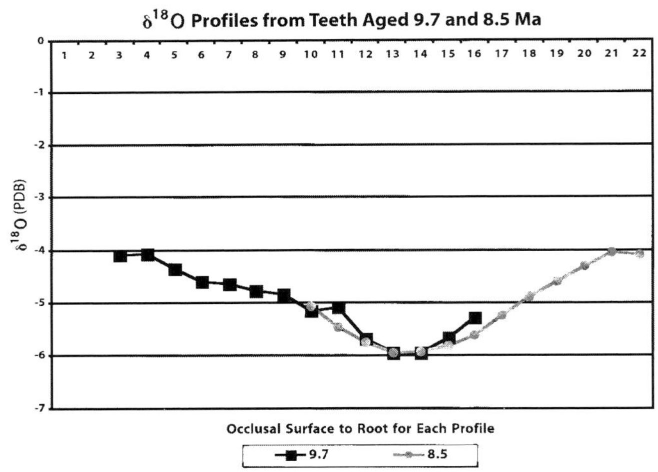

Figure 4.24 $\delta^{18}O$ values from two different teeth, aged 9.7 and 8.5 Ma, overlap for seven samples surrounding the trough, indicating that profiles representing different seasons can be combined to form the full annual cycle.

cycle over a sample size of only 38 mm, making the curve shorter from crown to root compared to other teeth. This may reflect greater time-averaging effects from sampling overlapping striae. If this tooth does not represent a full year, then it suggests a year characterized by multiple short rainy seasons. Alternatively, if this individual migrated over time, it would track precipitation regimes in multiple geographic regions and habitats rather than one region. The 8.0 Ma tooth was sampled over 53 mm and suggests a full seasonal cycle from wettest months to driest and falling back to wettest. However, while the second shift (from driest to wettest) shows an amplitude similar to other teeth, the first shift (from wettest to driest) shows a much smaller amplitude, suggesting the individual's water source was buffered at first or vegetation made a greater contribution to body water. This horse may have drunk from a buffered river or spring initially and switched to shallow, less buffered drinking holes later. Alternatively, the first rainy season represented by this tooth yielded less rainfall than typical years. The 7.9 Ma tooth (44 mm) is very similar in pattern to the 8.0 Ma tooth (see Figure 4.26). The 6.9 Ma tooth (46 mm) suggests a shift from wet to dry season followed by a return to wet, but the amplitude is again smaller than most other teeth. Again, this suggests a buffered water source, water from vegetation, or a rainy season with less overall rainfall.

Among the eleven teeth which did yield the consistent annual precipitation curve, all but the 9.0 Ma tooth group into three curves with respect to their $\delta^{18}O$ values and can therefore be combined to from three composite curves, one with depleted $\delta^{18}O$ values between about –4 and –6‰, one intermediate with values around –1 to –3‰, and one with values around +2 to 0‰ (see Figure 4.27). The most depleted curve consists of the oldest samples of 10.0 and 9.7 as well as an 8.5 Ma sample. The 8.3 Ma tooth with a complete cycle also falls with this composite curve in terms of $\delta^{18}O$ values. The middle curve consists of the 9.3, 8.7, and 8.4 Ma teeth. The most enriched curve consists of the youngest teeth, ages 7.7, 7.2, and 6.3 Ma. These results suggest a transition from closed habitat to more open habitat over time, accompanied by and associated with a decrease in rainfall over time, although the seasonal cycle itself, including its amplitude, does not appear to change. In each of

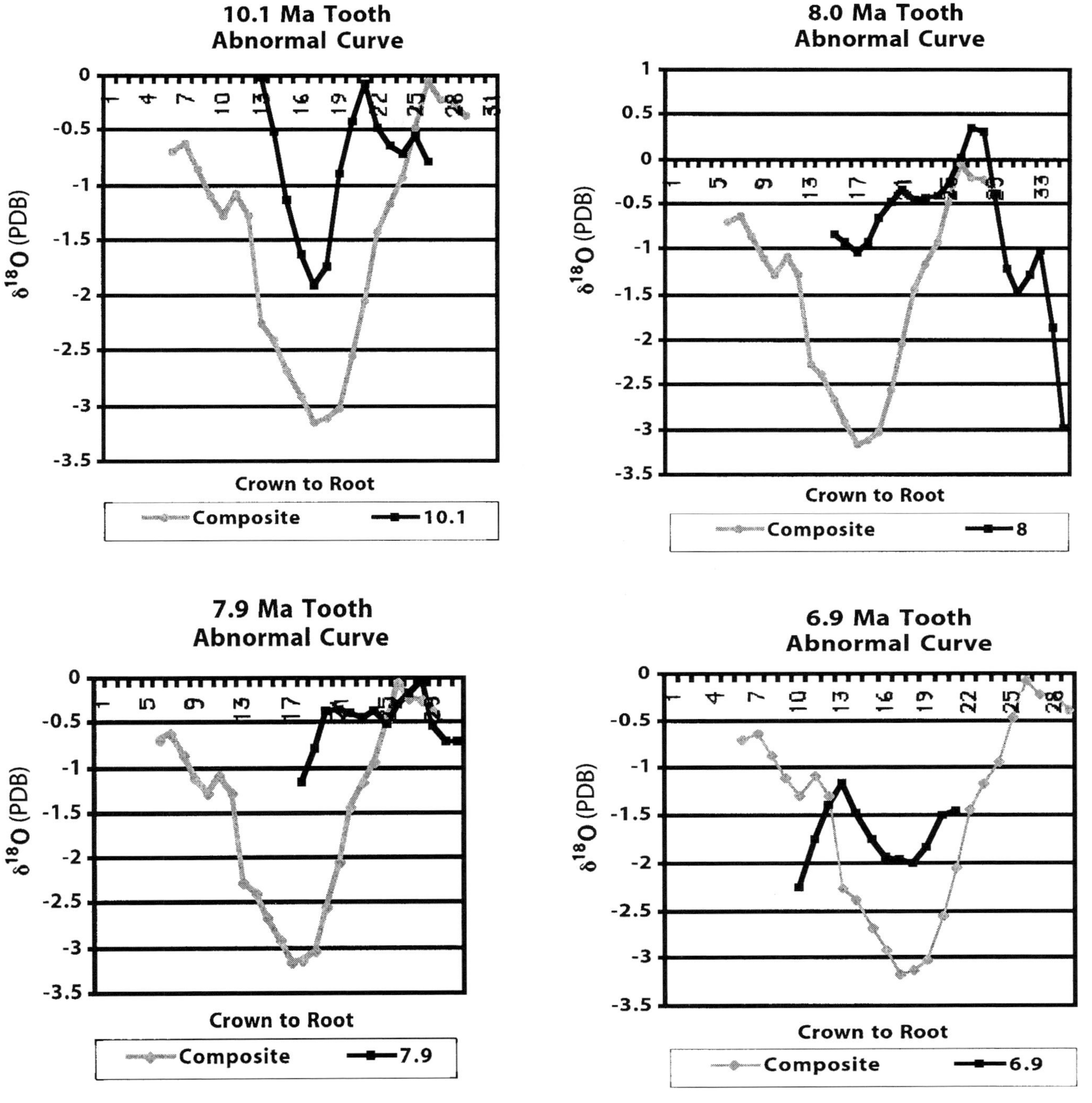

Figure 4.25 Four teeth yielded $\delta^{18}O$ curves which did not fit the shape of the curve generated by all other teeth (represented by the composite curve in each graph).

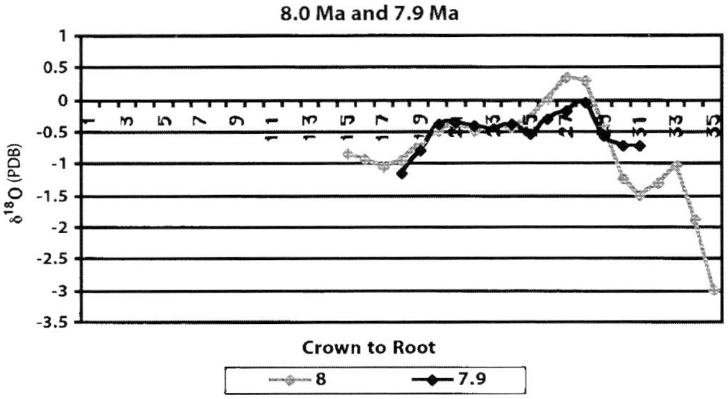

Figure 4.26 The 8.0 and 7.9 Ma teeth, both M2s, yielded abnormal curves with respect to the typical tooth profile, but they track one another.

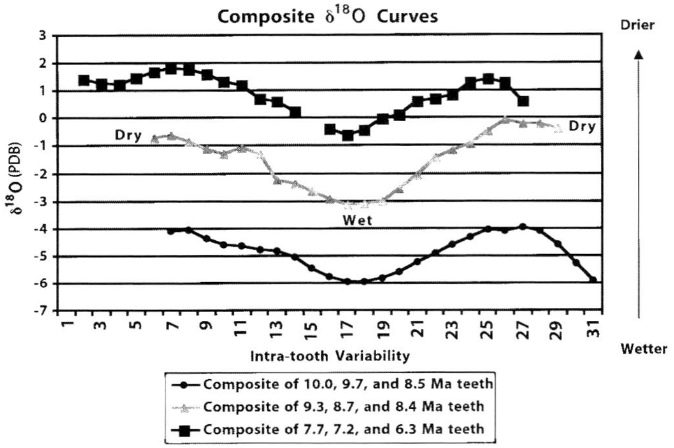

Figure 4.27 Overlapping tooth profiles can be combined to form three composite $\delta^{18}O$ curves. Oldest specimens comprise the most depleted curve, while youngest specimens comprise the most enriched curve. Teeth between 9.3 and 8.4 Ma fall into the two lower curves. While $\delta^{18}O$ values of curves become more enriched with time, suggesting a decrease in annual rainfall over time, the shape of the curve does not change, suggesting the seasonal regime remained the same.

these curves, the full seasonal cycle is represented by about 19 samples, with consecutive samples creating a similar pattern. When amplitude is measured for both the right and left sides of the trough and then averaged, the most depleted curve has an amplitude of 1.9‰; the middle curve, 2.9‰, and the enriched curve, 2.2‰, suggesting little to no change in the difference in amount of rainfall between dry and wet months.

The C and $\delta^{18}O$ values, and the amplitude of the $\delta^{18}O$ seasonal curves can be better interpreted when compared to Miocene *Hipparion* spp. from other geographic regions and to Siwalik bovid intra-jaw samples. Four Miocene *Hipparion* teeth from the 10.8 Ma Chorrora formation of Ethiopia and four *Hipparion* teeth from the 10.5 Ma Dinotheriesandes of Germany were sampled, with two teeth from each region sampled three times, once from the occlusal surface, once midway between the occlusal surface and root, and once from the cervical margin (see Table 4.5).

The C values for the Chorrora teeth are much more enriched than the oldest Siwalik teeth and represent the oldest evidence so far of C_4 graze, around two million years older than the first clear evidence of C_4 graze in the Siwaliks. Each of the four teeth indicates a mixed diet including C_4 graze, and in the two teeth with mini-profiles, C values correlate positively with $\delta^{18}O$ values, suggesting the C_4 component of the diet increases during the drier part of the year. This result should be interpreted with caution, however, for only two teeth are represented, and one of those teeth showed only a 0.5‰ variability in carbon values. The Dinotheriesandes teeth, on the other hand, are more depleted in C values than the oldest *Hipparion* spp. in the Siwaliks, yielding values consistently lower than –14‰. These depleted values are equal to the lowest values found among Siwalik fauna (see Chapter Three) and represent forested habitat, though not completely closed canopy.

The $\delta^{18}O$ values of the Chorrora teeth are as positive as and more positive than the youngest, most enriched teeth sampled from the Siwaliks, indicating both a different precipitation source and a more open habitat, consistent with the C results. Of the two teeth sampled multiple times, one tooth yielded a $\delta^{18}O$ range of 0.4‰ (sampled over 30 mm), while the other yielded a range of 3.2‰ (sampled over 29.8 mm). This discrepancy in ranges is not due to sampling length of the tooth and suggests that some buffering of water sources or enrichment from vegetation with respect to rainfall values has occurred, with heavy buffering in the case of the smallest range. The Dinotheriesandes teeth yielded $\delta^{18}O$ values similar to those of the depleted and middle composite Siwalik curves. Again, two teeth were sampled multiple times, with one tooth yielding a $\delta^{18}O$ range of 1.4‰ (over 31 mm of tooth) while the other had a range of 0.4‰ (over 38.4 mm of tooth), suggesting some buffering of water sources or enrichment of vegetation. In light of these discrepancies, the consistent amplitude of the Siwalik profiles suggests that less buffering took place in the Siwaliks, or alternatively, the Siwaliks may have experienced much more predictable seasonal changes in precipitation. Buffering may have been less of a factor in the Siwaliks compared to other regions for one of two reasons: 1) a greater presence of shallow bodies of water to serve as watering holes, or; 2) the monsoonal $\delta^{18}O$ signal is so strong that it masks any evaporative effects or contributions of groundwater to surface waters, as observed in north central African rivers (Fontes et al. 1970; Fritz 1981).

In addition to Miocene equid teeth from other regions, the Siwalik equid oxygen isotopic patterns can be compared to Siwalik bovids. Seven Siwalik bovid jaws representing *Dorcadoxa porrecticornis* and *Tragoceridus* spp. were sampled once on both an M2 and an M3 to yield ranges in $\delta^{13}C$ and $\delta^{18}O$ values (see Table 4.6). These jaws represent the two time levels, the S-level and the PS-level, analyzed in Chapter Three. *Dorcadoxa* M2-M3 pairs yielded $\delta^{18}O$ intra-tooth ranges between 3.2 and 5.2‰. *Tragoceridus* M2-M3 pairs yielded $\delta^{18}O$ ranges between 3.2 and 7.5‰. These results suggest that the amplitudes represented by the *Hipparion* profiles may be attenuated with respect to changes in isotopic values of rainfall, because greater ranges can be found in these bovids (by sampling multiple teeth). Another explanation is that in intra-tooth sampling, each sample may cover several striae of Retzius, with consecutive samples covering the same striae due to the angle of enamel deposition, thus yielding averaged $\delta^{18}O$ values rather than the true range. Fricke and O'Neil (1996) also found inter-tooth $\delta^{18}O$ variability to be greater than intra-tooth variability in a 500 year old bison jaw, with a 3.1‰ range found in eight samples from an M3, but 5.6‰ range between all premolars and molars of that jaw. They found a similar pattern in a modern sheep jaw, with intra-tooth variability of 2.3–2.5‰ but an intra-jaw variability of 3.8‰. These discrepancies could result from inter-year variability, for tooth development for the entire jaw in cattle today takes place over several years (Hillson 1986). They could also result from migrations in the bison, or alternatively, the smaller intra-tooth ranges could result from time-averaging because of sampling several striae laid down at different times.

Table 4.5

Miocene equids sampled from Chorrora and Dinotheriesandes

Specimen	$\delta^{13}C$	$\delta^{18}O$
Chorrora 3-crown	-5.7	0.6
Chorrora 3-middle	-6.1	0.9
Chorrora 3-root	-5.6	0.5
Chorrora 4	-5.1	0.9
Chorrora 2	-4.6	1.2
Chorrora 1-crown	-7.5	4.2
Chorrora 1-middle	-6.4	3.3
Chorrora 1-root	-5.0	1.0
Dino 3177-crown	-15.0	-3.2
Dino 3177-middle	-14.4	-2.8
Dino 3177-root	-14.2	-4.2
Dino 2763	-14.0	-3.4
Dino 2760	-14.1	-5.3
Dino 2659-crown	-14.7	-4.3
Dino 2659-middle	-14.0	-3.9
Dino 2659-root	-13.4	-4.2

Miocene *Hipparion* spp. teeth sampled from the 10.8 Ma Chorrora Formation in Ethiopia and the 10.5 Ma *Dinotheriesandes* of Germany. The Chorrora teeth yield the oldest evidence of C_4 graze so far. The *Dinotheriesandes* teeth yield depleted $\delta^{13}C$ (PDB) values indicative of a forested habitat. Both geographic regions yield teeth with differing intra-tooth ranges in $\delta^{18}O$ (PDB) values, suggesting some buffering of water sources relative to rainfall occurred, or alternatively, a less predictable seasonal rainfall regime than that of the Siwaliks, where $\delta^{18}O$ ranges were more consistent.

In addition to yielding greater ranges in oxygen values compared to individual equid teeth, these bovid intra-jaw samples differ from equids in showing consistent trends in $\delta^{13}C$ and $\delta^{18}O$ values from M2s to M3s.

Table 4.6

Intra-jaw profiles of Dorcadoxa and Tragoceridus (bovids)

Taxon	Specimen	Age (Ma)	M2 $\delta^{13}C$	M3 $\delta^{13}C$	M2 $\delta^{18}O$	M3 $\delta^{18}O$	Range $\delta^{18}O$
Dorcadoxa	Y2817	8.1-8.0	-10.9	-9.6	1.6	-3.6	5.2
Dorcadoxa	Y461	8.1-8.0	-11.9	-10.7	-3.2	-8.0	4.8
Dorcadoxa	Y9207	8.1-8.0	-10.3	-10.6	-1.1	-4.3	3.2
Tragoceridus	Y5127	9.3-9.2	-11.6	-11.5	-6.7	-3.5	3.2
Tragoceridus	Y11593	9.3-9.2	-11.7	-10.7	-4.4	0.7	5.1
Tragoceridus	Y9204	8.1-8.0	-11.0	-10.0	-7.0	0.5	7.5
Tragoceridus	Y263	8.1-8.0	-11.1	-9.8	-5.1	-1.3	3.8

Three *Dorcadoxa porrecticornis* jaws and four *Tragoceridus* spp. jaws were sampled once on both the M2 and M3. $\delta^{18}O$ (PDB) intra-jaw ranges are greater than the intra-tooth *Hipparion* spp. ranges. In *Dorcadoxa*, M2s are the most enriched in $\delta^{18}O$ values, while M3s are the most enriched in $\delta^{13}C$ (PDB) values. In *Tragoceridus* spp., M3s are the most enriched in both $\delta^{18}O$ and $\delta^{13}C$ values.

These trends differ between the two genera, however (see Figure 4.28). In *Dorcadoxa*, $\delta^{18}O$ values are always more enriched in M2s compared to M3s, while $\delta^{13}C$ values in two of the three jaws are more enriched in M3s rather than M2s. In the third *Dorcadoxa* jaw, carbon values are the same in both teeth. In *Tragoceridus*, on the other hand, both $\delta^{18}O$ and $\delta^{13}C$ values are enriched in M3s compared to M2s in three of the four jaws, with one jaw yielding similar carbon values for both teeth but the predicted pattern for oxygen values. The enrichment of carbon values from M2s to M3s, and hence over time for the two bovids indicates feeding in more open, more evaporative habitats as individuals develop. The discrepancy in oxygen trends can be interpreted in several ways. First, it is possible that the bovid taxa had different birthing seasons. Second, the reduncine *Dorcadoxa* is smaller than the boselaphine *Tragoceridus* and could have grown much faster. Alternatively, rather than indicating different seasons, the $\delta^{18}O$ values could indicate that *Tragoceridus* was always an obligate drinker, while *Dorcadoxa* was only an obligate drinker during the dry season, thus resulting in the negative correlation between carbon and oxygen values.

The depleted $\delta^{18}O$ values of the bovids suggest that the actual $\delta^{18}O$ values of the equid tooth profiles cannot be compared to modern values, for they are enriched due to evaporative effects. Faunal isotopic analyses presented in Chapter Three corroborate these results, for they showed that *Hipparion* was always in the most open, most evaporative environments relative to other taxa, with more enriched $\delta^{18}O$ values than most other taxa. Furthermore, the actual amplitude of the tooth profiles most likely represents a minimum amplitude of the precipitation $\delta^{18}O$ values rather than the full amplitude, for some buffering of water sources, some buffering due to vegetation contributions to body water, and possibly some time-averaging due to sampling multiple striae may have occurred. However, the fact that the shape of the seasonal curve does not change throughout the record, including the amplitude and the distance between highest and lowest values (measured by number of samples), suggests that the shape of the curve can be compared to modern precipitation $\delta^{18}O$ profiles.

The International Atomic Energy Agency (IAEA 1981; IAEA/WMO 2001) has collected monthly precipitation amounts and $\delta^{18}O$ values (weighted for amount) of precipitation for many locations around the world. These modern profiles include an African city in a region of evergreen and semi-evergreen rainforests characterized by a bimodal distribution of rainfall (see Figure 4.29), southeast Asian cities in evergreen rainforest regions with greater than 100 mm rainfall every month of the year (see Figure 4.30), southeast Asian cities in monsoonal forest regions with distinct wet/ dry months due to the Asian monsoon (see Figure 4.31 and 4.32), and an Indian city which also experiences a monsoonal climate, but with more dry months than wet per year (see Figure 4.33).

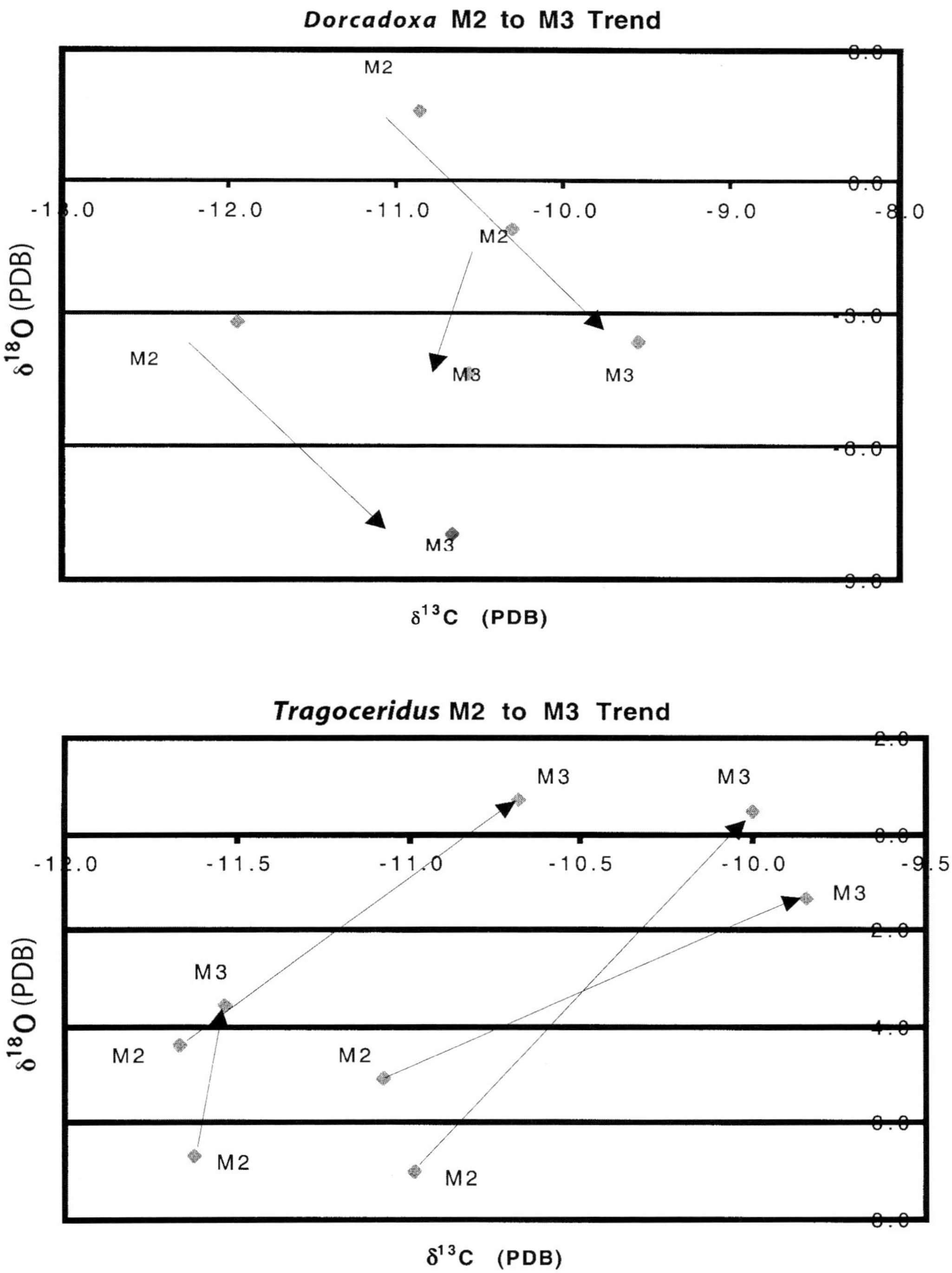

Figure 4.28 *Dorcadoxa* and *Tragoceridus* exhibit different trends in $\delta^{18}O$ values between M2s and M3s, and hence, over time. Arrows are drawn from M2s to M3s of the same jaw. *Dorcadoxa* becomes more depleted; *Tragoceridus*, more enriched.

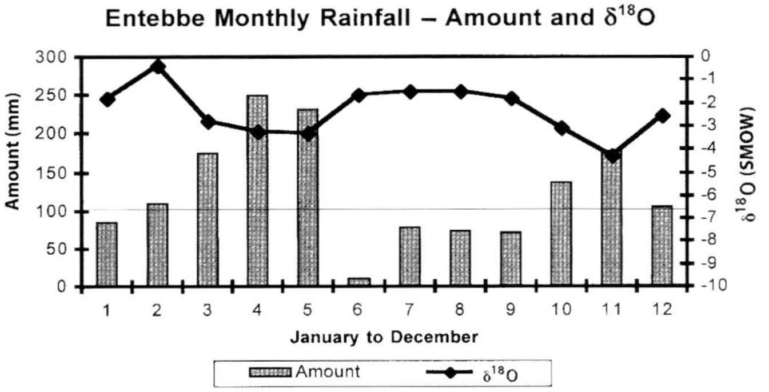

Figure 4.29 Typical monthly rainfall amount and δ18O (SMOW) values collected by the IAEA (1981). Entebbe, Uganda, represents the bimodal annual rainfall regime of East Africa. There are two drops in δ18O values which correspond to the two rainy seasons.

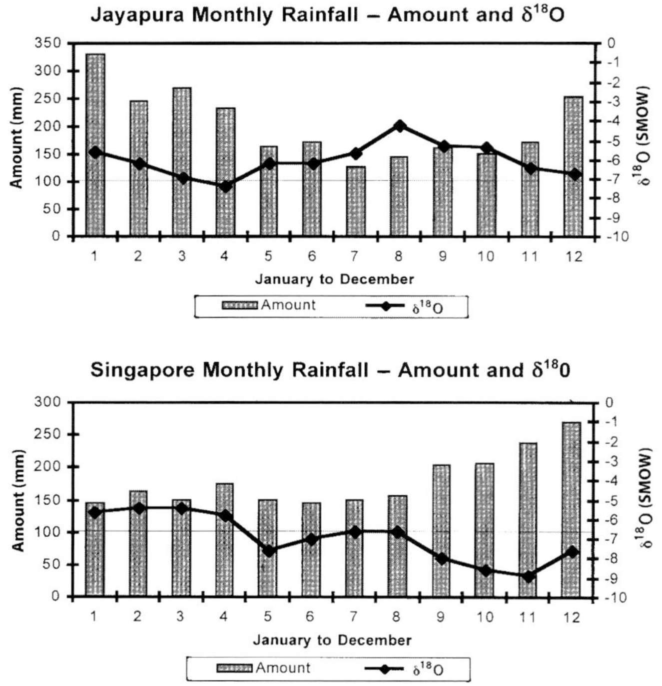

Figure 4.30 Jayapure, New Guinea, and Singapore represent two East Asian regions in which monthly rainfall amounts never fall below 100 m (IAEA 1981). While δ18O (SMOW) values vary, they do not follow a distinct seasonal trend.

The first thing to note is that the Siwalik precipitation curve is not bimodal and is therefore more similar to Asian curves. Among the Asian curves, those in regions which never experience dry months (months receiving less than 100 mm of rainfall) show variability in δ18O values, but this variability does not show clear seasonal trends. Jakarta, which experiences four months of rainfall just under 100 mm, shows a more seasonal trend in δ18O values. However, cities in monsoonal regions which experience four-to-five considerably dry months per year show a distinct seasonal difference in δ18O values which results in a U-shaped curve. There is a small plateau of enriched values representing dry months, followed by a gradual decline in values as rainfall increases, which represents the onset of the monsoon. This decline ends in a trough which represents several months of extremely heavy rainfall, followed by a gradual rise in δ18O values representing decrease in rainfall as the monsoon recedes. Finally, in the monsoonal environment characterized by more dry months than wet, the precipitation isotopic curve also indicates seasonal transitions, though different from the wet monsoonal curves. In New Delhi, which experiences around nine dry months per year, there is a long plateau of enriched values representing dry months, followed by a very sharp drop to depleted values, representing the onset of the monsoon. These depleted values last only a few months, however, before the curve rises very sharply, representing a return to dry months as the monsoon recedes.

The Siwalik precipitation curve is U-shaped, with small, enriched plateaus, a gradual increase in precipitation, a short trough, and a gradual rise in values. This curve looks very much like the wet monsoonal environments of southern China today. It does not look like the curves which result from heavy rainfall year-round, nor does it resemble curves resulting from annual rainfall regimes characterized by more dry months than wet (see Figure 4.34). Furthermore, it reflects a seasonal trend which characterized places with months considerably drier than 100 mm rainfall and with dry periods of five-to-six months (unlike Jakarta, where there is a four month dry season, but rainfall is only just under 100 mm per month).

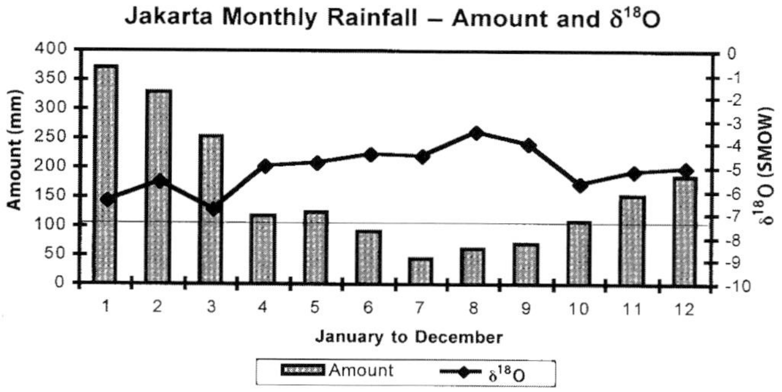

Figure 4.31 Monthly rainfall amount and $\delta^{18}O$ (SMOW) values for Jakarta, Indonesia (IAEA 1981). Jakarta experiences several months with less than 100 mm per year. $\delta^{18}O$ values become enriched during the drier months, but the overall pattern is largely a plateau, like wetter regions of East Asia.

These results suggest seasons of very heavy rainfall in contrast to seasons of considerably less rainfall, or the pattern of monsoonal climates today. These results therefore suggest that the Asian monsoon was probably already in place by 10.0 Ma, and that throughout the record from 10.0–6.3 Ma the Siwaliks experienced heavy monsoonal rainfall for at least several months, with five-to-six months of dry season per year. The fact that some teeth incorporated only the dry to wet transition, while others incorporated both the full trough and the rise to the dry season suggests that the transition from dry to wet took a longer period of time than the transition of wet to dry. This discrepancy could represent similarity to the timing of the monsoon in India today, where onset of the monsoon is gradual, but recedes very quickly (Mooley and Shukla 1987). This pattern of gradual onset but rapid departure is likely the result of soil moisture effects. As land becomes warmer than the ocean, moisture-laden air masses are drawn to land. The more it rains in any particular patch of land, however, the more moist the soil becomes, and higher its specific heat index becomes, or the greater an amount of heat is needed to raise the soil temperature by one degree Celsius (Webster 1987). As soils become soaked, neighboring dry land becomes warmer than wet land, resulting in movement of the air mass to the neighboring patch of dry land. This process continues until the ocean becomes warmer than land once again. Now, soil moisture has much less of an effect, for the air mass is attracted to the far ocean rather than to neighboring patches of land, resulting in a very rapid departure.

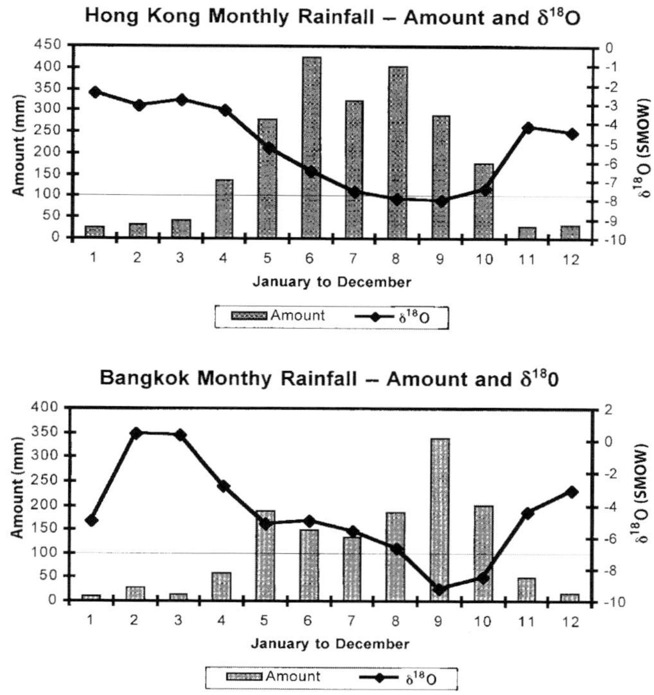

Figure 4.32 Hong Kong and Bangkok experience a very monsoonal rainfall regime, with 5-6 months of little rainfall, and the remaining months experiencing very heavy rainfall. $\delta^{18}O$ (SMOW) values show a U-shaped trend (IAEA/WMO 2001; IAEA 1981).

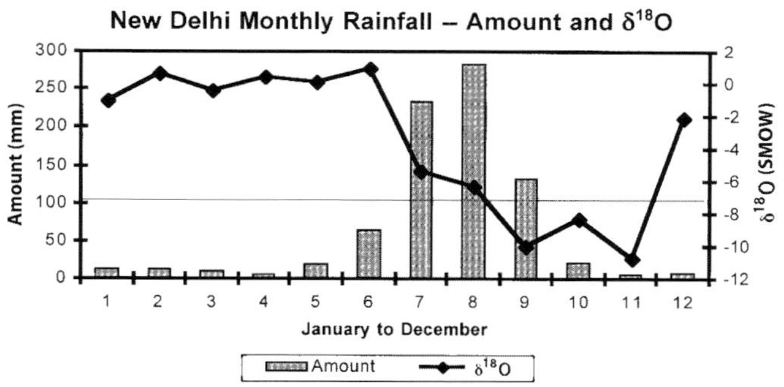

Figure 4.33 New Delhi, India experiences the Asian monsoon, with several months of very heavy rainfall, but nine months of very little rainfall. Consequently, $\delta^{18}O$ (SMOW) values show a plateau of enriched values, followed by a very sharp drop, followed by a very sharp rise (IAEA 1981).

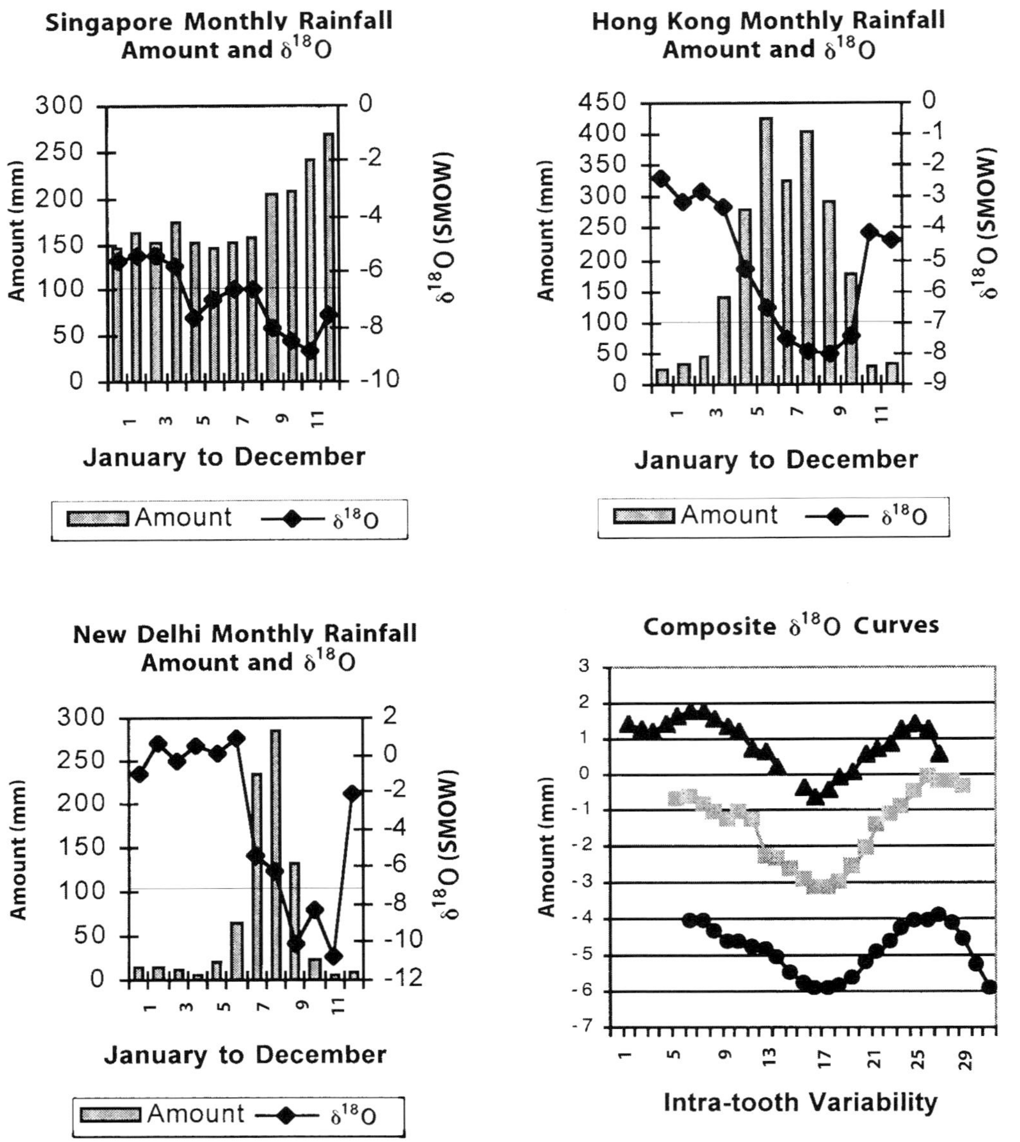

Figure 4.34 The Siwalik $\delta^{18}O$ curve most resembles Hong Kong, representing a monsoonal rainfall regime with five-to-six months dry season and several months of very heavy rainfall.

The overall pattern from 10.0–6.3 Ma of increasing δ18O values over time but with maintenance of the same magnitude of seasonal variability (amplitude) is the same as the pattern yielded by Nepal bivalve shells spanning 9.5 to 3.1 Ma (Dettman et al. 2001). Increasing δ18O values suggest a decrease in amount of annual rainfall over time, but no change in curve shape suggests no change in rainfall regime with respect to number of dry versus wet months.

These results suggest that *Sivapithecus* lived under a more seasonal rainfall regime than any great ape experiences today, with the exception of the Mt. Asserik chimpanzees (Baldwin et al. 1982; McGrew et al. 1981). Only a few gibbon populations are able to inhabit monsoonal forests today (Chivers 1984; Kappeler 1984). These results therefore suggest that either *Sivapithecus* had habitat requirements different from those of modern apes, or alternatively, Siwalik forests under a monsoonal rainfall regime differed in the spatio-temporal availability of fruit compared to modern analogues, thus providing habitat that could support a great ape. These alternative scenarios will be explored further in Chapter Five.

Eleven of fifteen teeth yielded the same seasonal curve, suggesting that not only is the curve an accurate representation of the rainfall regime, but that the rainfall regime was generally highly predictable, a further indication of a monsoonal environment. The four "abnormal" teeth indicate some sampling time-averaging, as well as buffered water sources, which resulted in teeth that did not yield accurate representations of the annual rain cycle.

If rainfall regime was in fact predictable, and if resource availability tracked rainfall, seasonal birthing could track them both. The horse teeth sampled in this study are worn, preventing estimates of crown formation time, but timing of when crown formation stops can be estimated from the final samples from each tooth, those taken at the cervical margin. A caveat is that any differences in hypsodonty, or tooth length, between specimens representing different species could result in longer periods of tooth development and hence different periods of crown completion even though the timing of initiation of crown formation may have been the same. The sample from this study includes three P3s, two P4s, 3 M1s, two M2s, and five M3s. When final samples are compared per tooth type, a seasonal signal is apparent, with

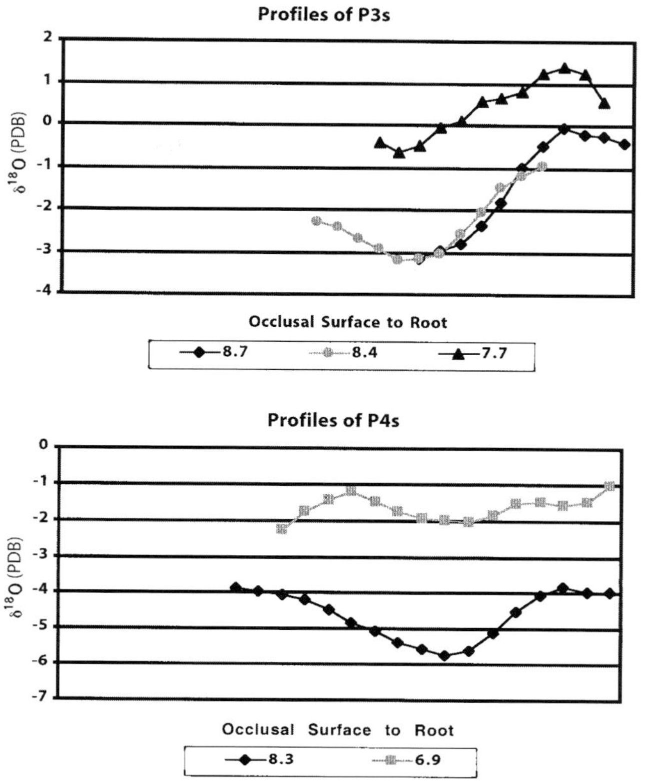

Figure 4.35 Profiles of P3s and P4s end around the crest, or most enriched δ18O values, indicating crown completion during the dry season.

crown termination ending during particular seasons for each tooth position, although termination time is not completely consistent among the M1s (see Figures 4.35 and 4.36). All P3s end as the teeth were approaching the crest or just leaving the crest, indicating crown formation stopped during the dry season. Among P4s, one of the two teeth is an abnormal curve, but both end with the highest δ18O values for those teeth, also indicating crown completion during the dry season. Among M1s, two curves end around the trough, or wet season, while one clearly stops formation during the dry season. Two teeth with abnormal curves represent the M2s, but they track one another (see Figure 4.26), suggesting similar timing of development, and they end approaching the trough, suggesting the onset of the wet season. Finally, among the M3s, two curves end approaching the trough (early wettest season), two end just leaving the trough (late wettest season), and one tooth is again an abnormal curve.

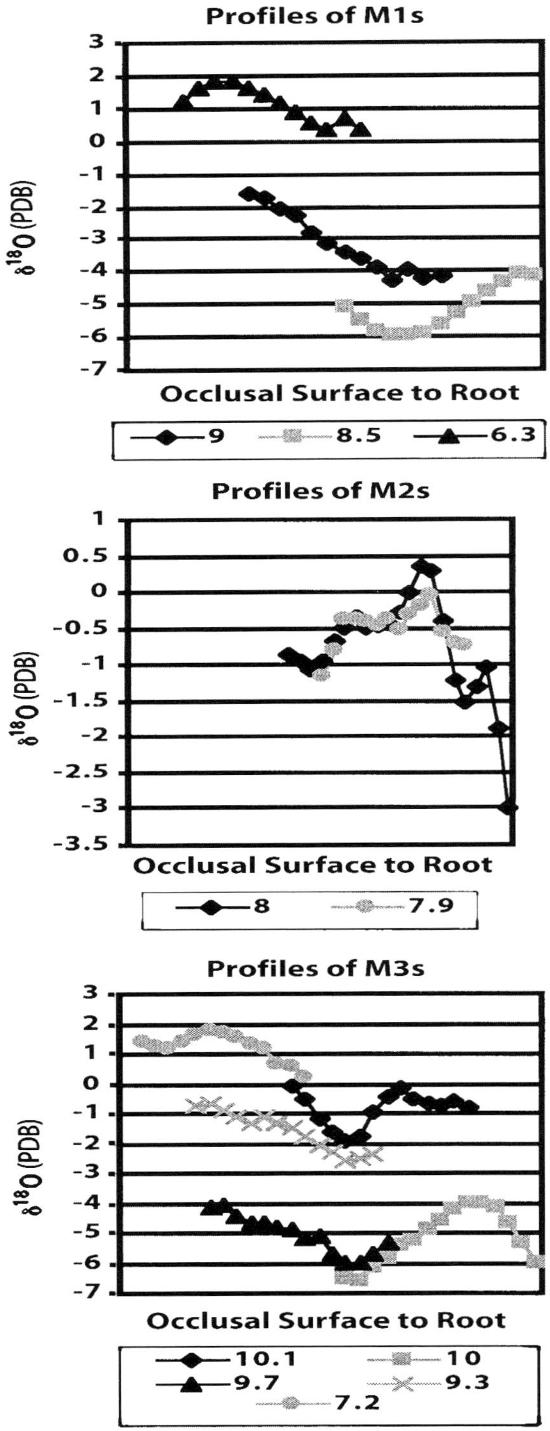

Figure 4.36 Profiles of M1s, M2s, and M3s almost all yield curves which end around the trough, or most depleted $\delta^{18}O$ values, suggesting crown completion during the wet season. One M1 indicates crown completion during the dry season.

Among Burchell's zebra today, births can occur in any month, but most births occur during the wet season (Nowack 1999). The Siwalik teeth are compatible with this flexible pattern of seasonal birthing. The abnormal M1 tooth suggests a different season of development than all other teeth, but there could have been a difference in hypsodonty undetectable in worn teeth. Alternatively, that M1 could have been misidentified to tooth position, but the only other tooth position possible is an M2, and it does not fit the pattern of M2s either. Finally, Hillson (1986) reports crown formation times in cattle, which are also hypsodont and act as a good proxy for horses. In cattle, crown completion occurs between birth and 0.5 years for M1, 0.5–1 year for M2, 1–1.5 years for P3, and 1.5–2 years for P4s and M3s. The Siwalik teeth follow this pattern when mapped onto a two-year seasonal curve, and depleted $\delta^{18}O$ values at crown completion time in M1s suggest most births take place around the wet season (see Figure 4.37).

The pattern of tooth development suggested by isotopic analyses can be compared to two Siwalik fossil juvenile equid maxilla represented by premolars and molars with teeth in different stages of development. In specimen Y49606 from locality Y894 dated to 9.6 Ma, only the M1 and M2 have erupted. P2 and M3 are close to the occlusal surface, but P3 and P4 are far from the occlusal surface. Crown formation has completed in all teeth except the M3. Roots have formed only in the M1. This pattern of development suggests that M1 is completed first, followed by M2, with P2 the next tooth to develop. P3 and P4 appear to develop around the same time, after P2 development, with M3, the only tooth with the crown not yet completed, being the last tooth to develop.

Likewise, specimen Y262 from locality Y017 dated to 8.0 Ma, is represented by deciduous P2, P3, and P4 caps, all occluded, and by the M1 and M2. The M1 has erupted, but roots are not yet complete. M2 has not yet reached the occlusal surface. This specimen, like the previous specimen, suggests that M1 is the first tooth to complete development, followed by M2, with premolars developing later. Both specimens are compatible with the tooth development sequence suggested by the isotopic analyses.

Conclusions

Carbon values for *Hipparion* teeth spanning 10.3–6.3 Ma indicate the first evidence for C_4 graze in the Siwaliks at 8.7 Ma. The first evidence, so far, of C_4 graze anywhere was found in the 10.8 Ma Chorrora teeth from Ethiopia. Though C_4 appears in the diet at 8.7 Ma in the Siwaliks, it does not appear consistently until 8.3 Ma. Even then, profiles still indicate some periods within a tooth where there is no C_4 component. However, by 7.9 Ma, C_4 is a component throughout tooth profiles, suggesting the first evidence for more extensive permanent C_4 grass patches. This result is compatible with the paleosol data showing the first C_4 grasses at 8.0 Ma (Quade and Cerling 1995; Quade et al. 1989). Intra-tooth profiles indicate that while C_4 graze became a component of *Hipparion* diets, these horses were eating a mixed diet that included browse throughout the record sampled (until 6.3 Ma). There is no seasonal trend in C_4 grazing or open/closed habitat feeding among the horses sampled. However, multiple samples for bovid individuals suggested shifts in habitat or feeding and drinking behavior throughout tooth development. These shifts appear to differ between the two bovid species sampled.

Eleven of fifteen tooth profiles indicate that intra-tooth sampling of hypsodont teeth can yield accurate representations of the seasonal trends in precipitation. This study indicates that in worn teeth, a greater length of enamel is needed than was suggested by original studies, greater, in fact, than most worn teeth permit. However, by sampling multiple teeth, which includes an array of premolars and molars, different portions of the annual seasonal signal can be pieced together. Furthermore, this study suggests that there is no weaning effect and all molars can be analyzed.

Single samples taken from horse teeth spanning 10.3 to 6.3 Ma, in addition to average C and $\delta^{18}O$ values from tooth profiles, indicate a shift to more open habitat and less annual rainfall over time. When specimens are divided into those that lived prior to the extinction of *Sivapithecus* and those that lived afterwards, the shift in oxygen is 2.1‰, compatible with the 1.8‰ shift observed among all fauna, reported in Chapter Three. However, these shifts include teeth representing all seasons of the year, and profiles from individual teeth

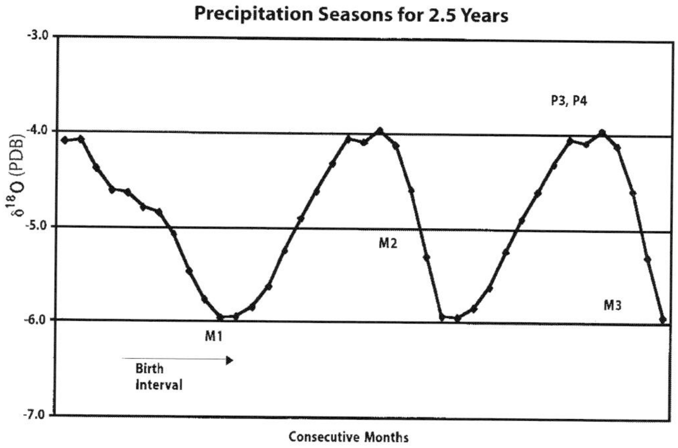

Figure 4.37 Timing of crown completion can be inferred from tooth profiles, and when mapped onto a 2.5 year seasonal curve generated from replicating a composite curve, suggests birthing in the wet season for *Hipparion* with a tooth development trajectory similar to but possibly longer than that of modern cattle (also hypsodont).

indicate that these seasons span at least 2‰, given buffering effects. A more accurate interpretation of change in amount annual rainfall can be obtained from comparing $\delta^{18}O$ values of the wettest months and again for the driest months between the most depleted composite curve, representing the oldest samples, and the most enriched composite curve, representing the youngest samples. These differences are 5.3 and 5.8‰ respectively. An average increase of 5.6‰ can be translated to a decrease in annual rainfall using the Dansgaard (1964) relationship described in Chapter Three, which indicates a −1.5‰ shift with every 100 mm of rainfall added in the tropics. This relationship suggests that over the time span of 10.0–6.3 Ma, rainfall in the Siwaliks decreased by approximately 375 mm.

This decrease in rainfall can be interpreted by comparisons of different ecosystems in India today, with data collected by Walter (1973). Belgaum, a semi-evergreen forest, experiences four dry months per year with an annual rainfall of 1,293 mm. Bidar, a moist-monsoon forest, experiences six dry months with a total of 858 mm annual rainfall. Jaipur, a dry-monsoon forest, has seven dry months and 610 mm of rainfall per year. Thus, 435 mm of rainfall and one extra dry month separates a semi-evergreen forest from a moist-monsoon forest; 248 mm and an extra dry month separate moist- and dry-monsoon forests. In a compilation of many Indian forests, for those which experience five month dry periods, 700 mm separate semi-evergreen from moist-monsoon forest, and about 300 mm separate moist- from dry-monsoon forest. For those forests which experience six month dry periods, 500 mm separate semi-evergreen and moist-monsoon forests; 400 mm separate moist- and dry-monsoon forest. Savannas in India occur only in regions where at least seven months of the year are dry, and those regions receive on average several hundred millimeters less rainfall than dry-monsoonal forests.

The seasonal precipitation curve interpreted from the tooth profiles most resembles rainfall regimes of southern China today, where monsoonal forests occur. These regions typically experience five-to-six month dry seasons. The decrease in rainfall suggested by the enrichment of $\delta^{18}O$ values over time in the composite curves was sufficient, according to modern Indian forest habitats experiencing similar dry periods, to drive a transition in vegetation from moist- to dry-monsoonal forest over time.

The seasonal regime of all the teeth profiled indicate that onset of the Asian monsoon likely occurred prior to 10.0 Ma. Furthermore, this precipitation regime is more seasonal than any great ape experiences today, with the exception of the Mt. Asserik chimpanzees. Therefore, either *Sivapithecus* differed in habitat and feeding requirements from modern apes, or the Miocene forests differed from modern forests in some way that allowed for less variable fruit availability (to be explored further in Chapter Five). A further indication of a monsoonal environment, or one with highly predictable seasons, is a seasonal birthing signal suggested by different tooth positions among the profiles, with most births occurring in the wet season, as African zebras today.

While the $\delta^{18}O$ values of the profiles indicate open/closed habitat as well as a decrease in rainfall over time, the youngest curve to fall among the most depleted values occurs at 8.3 Ma, suggesting loss of closed habitat around the time *Sivapithecus* went extinct (8.4 Ma) as well as an initial decrease in rainfall. Furthermore, all three samples at 7.7 Ma or younger make up the most enriched profiles, indicating a second transition well after *Sivapithecus* went extinct, most likely suggesting a transition from wet to dry monsoonal forest accompanied by a decrease in rainfall, with this transition occurring around the time that paleosol carbon isotopes detect the first C_4 grasses. This transition also coincides with the 7.5 Ma decrease in rainfall detected in the Nepal bivalve shells (Dettman et al. 2001).

5 Synthesis and Conclusions

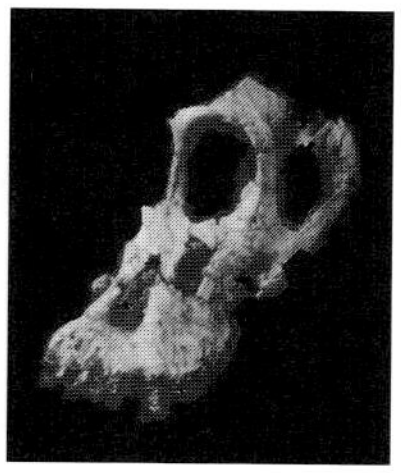

Synthesis of Microwear and Isotopic Analyses

This research adopted three approaches to making inferences concerning fruit availability, the vegetation mosaic, and the annual rainfall regime throughout the Late Miocene as recorded in the Siwaliks and particularly for *Sivapithecus* habitats. The first presented dental microwear analyses of a wide range of Siwalik mammals. This study yielded evidence for frugivory among *Sivapithecus*, suids, tragulids, and some bovids, and results for some frugivorous taxa suggest periods or geographic regions of low fruit availability, and a decrease in availability over time. The second study presented isotopic analyses of numerous mammalian taxa for two time intervals bracketing the extinction of *Sivapithecus*. These analyses provided vegetation mosaic reconstructions with different species exploiting different parts of the habitat. A comparison of the two intervals suggests that the mosaic changed over time, with forests being replaced, but not eliminated, by more open habitat, including patches of C_4 grass, by 8.1 Ma. Finally, the third study involved isotopic analyses of intra-tooth and intra-jaw samples for equids and some bovid individuals. This analysis yielded evidence for a seasonal rainfall regime throughout the Late Miocene that was similar to monsoonal rainfall patterns experienced today in southern China. It also provided the oldest indisputable evidence for C_4 graze in the Siwaliks at 8.7 Ma. Intra-tooth sampling of equid specimens indicated that graze, once incorporated into the diet, was a consistent element year-round. Multiple samples for bovid individuals suggested shifts in habitat or feeding and drinking behavior throughout tooth development. Finally, isotopic patterns among different premolars and molars suggest seasonal birthing for equids, with most births occurring during the wet season.

Each of these studies overlapped in terms of the species analyzed. As suggested in the dental microwear analyses of Chapter Two, dietary and habitat preferences for each species are best reconstructed by incorporating both microwear and isotopic analyses. Without corroborative isotopic analyses, microwear resulting from hard-shelled fruits and seeds could not be distinguished from wear resulting from C_4 phytoliths. On the other hand, while isotopic analyses provide quantifiable differences between feeding on C_3 versus C_4 plants and between feeding in closed versus open habitat, they do not allow finer distinctions between C_3 browsers and C_3 frugivores or between C_3 browsers and C_3 grazers. A synthesis of all lines of evidence for each taxon will yield better dietary and habitat reconstructions for each taxon, which are essential for placing the extinction of *Sivapithecus* in a context of faunal turnover. The following is a list of all taxa included in this thesis, with results from isotope and/or microwear analyses reported. For some taxa, individual specimens were sampled both for isotopes and microwear (see Table 5.1). Correlations found between isotope and microwear results for individual specimens are noted below per species, but where no correlations are found, this is not unexpected given the methods are sampling from very different time periods in an animal's life. Isotopes sample the diet during tooth enamel formation, while microwear samples the last few days before death. Correlations between results of the two methods are found at the *taxon* level, however, as reported below. For most taxa, in addition to isotope and microwear results, first and last occurrences (oldest and youngest fossils) are reported, with first and last appearances (time range extensions, or confidence intervals, based upon a statistical method by Koch [1987]) in parentheses, thus demonstrating clustering of extinctions over time (dates from Barry et al. 2002).

Finally, Cerling et al. (2003) have built an isotopic library for east African bovids from which they suggest quantifiable $\delta^{13}C$ differences between different food preferences. Their dietary categories include feeding under

Table 5.1

Specimens sampled in both microwear and isotopic analyses

Specimen	$\delta^{13}C$ (PDB)	$\delta^{18}O$ (PDB)	Microwear
Propotamochoerus			
Y14502 (9.3 Ma)	-11	-6.9	Pot (.61) Bab (.34) Hylo (.05)
Y4554 (9.2 Ma)	-12.6	-5.4	Pot (.81) Bab (.18) Hylo (.02)
Y49515 (8.0 Ma)	-10.2	-3.6	Pot (.79) Bab (.12) Hylo (.09)
Y10224 (9.3 Ma)	-12	-8.4	Pot (.45) Bab (.09) Hylo (.46)
Y11007 (9.2 ma)	-11.2	-8.2	Pot (.41) Bab (.17) Hylo (.43)
Y41368 (9.3 Ma)	-12.9	-9.2	Pot (.24) Bab (.08) Hylo (.68)
Hippopotamodon			
Y5276 (9.2 Ma)	-9.4	-7	Pot (.37) Bab (.52) Hylo (.11)
Y27915 (9.2 Ma)	-11.2	-5.7	Pot (.62) Bab (.26) Hylo (.13)
Y13524 (9.2 Ma)	-10.8	-7.3	Pot (.45) Bab (.2) Hylo (.35)
Y330 (8.0 Ma)	-11.1	-4.1	Pot (.56) Bab (.05) Hylo (.39)
Dorcabune nagrii			
Y6857 (9.3 Ma)	-10.6	-2.4	Browse (.05)
Y10220 (9.3 Ma)	-11.1	-8.2	Browse (.03)
Dorcatherium majus			
Y5567 (9.3 Ma)	-11.1	-5.6	Browse (.07)
Y16236 (8.0 Ma)	-10.6	-1.7	Browse (.04)
Y49530 (8.0 Ma)	-9.9	-2.4	Browse (.47)
Tragoceridus			
Y5127 (9.2 Ma)	-11.6;-11.5	-6.7;-3.5	Br (.17) Gr (.05) M (.42) Fr (.33)
Y11593 (9.3 ma)	-11.7;-10.7	-4.4;+ 0.7	Br (.21) Gr (.15) M (.51) Fr (.13)
Dorcadoxa			
Y2817 (8.0 Ma)	-10.9;-9.6	+1.6;-3.6	Br (.81) Gr (.01) M (.16) Fr (.04)

Specimens sampled in both isotopic and microwear analyses, with ages in parentheses. These analyses sample different times in the animal's life, with isotopes sampling time during enamel formation, and microwear sampling the week before death. Nonetheless, among *Propotamochoerus* specimens, those for which microwear suggests a heavy fruit component, or diet more similar to African bushpigs (frugivores), are more enriched in $\delta^{18}C$ values compared to those specimens with microwear indicating more browse in the diet, or those specimens with microwear similarities to both bush pigs and giant forest hogs (folivores). Among other taxa, carbon isotopic results indicate that any heavy wear features resulting in grazing and mixed diet probabilities are most likely a result of feeding on hard fruit, for the isotopic signal indicates no significant amount of C_4 in the diet.

closed canopy, such as in the Ituri Forest, hyperbrowsers/frugivores that eat 100 percent C_3 plants, browsers that consume less than 30 percent C_4, mixed feeders that eat at least 30 percent of both C_3 and C_4, grazers feeding on at least 70 percent C_4, and hypergrazers feeding on 100 percent C_4. While their isotopic cutoffs represent a gradient of C_4 dietary composition, they also represent a gradient of closed to open habitat. In this Siwalik analysis, their cutoff values are adjusted by 1.5‰ to correct for the shift in isotopic composition of the atmosphere over the past 150 years due to the burning of fossil fuels. All taxa sampled isotopically are then plotted against their gradient, with each taxon represented by its mean and one standard deviation (see Figures 5.1 and 5.2).

Sivapithecus spp.: 12.3–8.5 Ma (14.0–8.4 Ma) Hominoidea

Isotopes

Sivapithecus is one of the most depleted species in carbon values for the S-level, with an average $\delta^{13}C$ value of −13‰, suggesting that it fed in closed forest. These values fall within the range of hyperbrowsers/frugivores (see Figure 5.1). Its oxygen values are very enriched, suggesting that *Sivapithecus* obtained most of its water from upper canopy vegetation.

Microwear

Microwear is characterized by numerous pits and coarse scratches, resembling all frugivorous large apes, with greatest similarity to orangutans. These results suggest

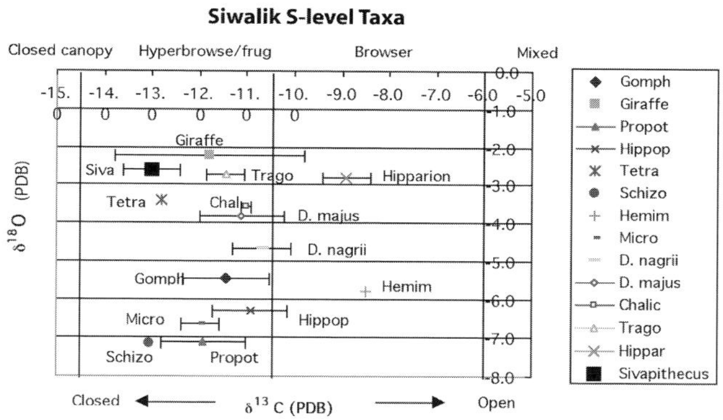

Figure 5.1 Cerling et al. (ms) suggest a gradient of $\delta^{13}C$ values that reflect different dietary categories in addition to a gradient of closed to open habitats. These categories are based on isotopic values of extant east African bovids. Their cutoff values are adjusted in this Siwalik analysis by 1.5‰ to correct for the shift in the isotopic composition of the atmosphere over the past 150 years due to burning of fossil fuels. The cutoff values thus become −14.5‰ or lower for feeding in closed canopy forest, such as the Ituri Forest; -14.5 to −10.5‰ for hyperbrowsers and frugivores (100% C_3); −10.5 to −6‰ for browsers (less than 30% C_4 graze), −6 to −2 for mixed feeders (greater than 30% for both C_3 and C_4 plants); -2 to +2.5 for grazers (greater than 70% C_4); and +2.5 and greater for hypergrazers (100% C_4). Siwalik taxa analyzed for the S-level include *Sivapithecus*, the suoids *Schizochoerus gandakasensis*, *Propotamochoerus hysudricus*, *Tetraconodon magnus*, and *Hippopotamodon sivalense*; the anthracotheres *Microbunodon punjabiense* and *Hemimeryx* lg. sp.; gomphotheres; the tragulids *Dorcabune nagrii* and *Dorcatherium majus*; the bovid *Tragoceridus* spp.; a chalicothere; a giraffe *Bramatherium megacephalum*, and the equid *Hipparion*.

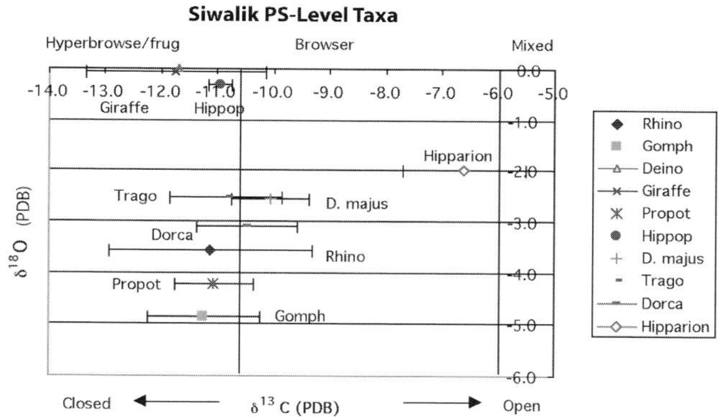

Figure 5.2 Cerling et al. (ms) suggest a gradient of $\delta^{13}C$ values that reflect different dietary categories in addition to a gradient of closed to open habitats. These categories are based on isotopic values of extant east African bovids. Their cutoff values are adjusted in this Siwalik analysis by 1.5‰ to correct for the shift in the isotopic composition of the atmosphere over the past 150 years due to burning of fossil fuels. The cutoff values thus become −14.5‰ or lower for feeding in closed canopy forest, such as the Ituri Forest; −14.5 to −10.5‰ for hyperbrowsers and frugivores (100% C_3); −10.5 to −6‰ for browsers (less than 30% C_4 graze), −6 to −2 for mixed feeders (greater than 30% for both C_3 and C_4 plants); −2 to +2.5 for grazers (greater than 70% C_4); and +2.5 and greater for hypergrazers (100% C_4). Siwalik taxa analyzed for the PS-level include gomphotheres; the suids *Propotamochoerus hysudricus* and *Hippopotamodon sivalense*; the tragulid *Dorcatherium majus*; the bovids *Tragoceridus* spp. and *Dorcadoxa porrecticornis*; a proboscidean *Deinotherium* sp.; a giraffid *Bramatherium* megacephalum, and the equid *Hipparion*.

that *Sivapithecus* was as frugivorous as most modern great apes, with some hard-object feeding, such as fruit with hard exocarps or hard seeds. Together, microwear and isotopic evidence indicates that *Sivapithecus* fed primarily on upper canopy fruits.

Propotamochoerus hysudricus: 10.2–6.8 Ma (10.2–6.5 Ma) Suinae

Isotopes

Propotamochoerus is one of the most depleted species in both $\delta^{13}C$ and $\delta^{18}O$ values for both the S- an PS-levels, suggesting it fed in a relatively closed, wet habitat in the vegetation mosaic. Among the six S-level specimens, half show carbon values of –12‰ or lower, indicative of forest. None have values higher than –11‰, and therefore do not approach the cutoff of –10‰ for recognizing open habitat. All S-level specimens fall below the average $\delta^{18}O$ value of –5‰ for the S-level as well, with three of the six falling outside one standard deviation from the mean. Those three oxygen values are all lower than –8‰, suggesting either a greater dependence on drinking water or a preference for wetter habitat relative to most other taxa.

Five specimens were sampled that dated to the PS-level. Again, no $\delta^{13}C$ values fall above –10‰, with two values of –11.8‰, indicative of relatively closed habitat. All oxygen values fall below the mean of –2.7‰ for PS-level specimens in general. PS-level $\delta^{18}O$ values are significantly more enriched than S-level values, suggesting a change in climate, such as a decrease in annual rainfall, or a greater use of drier habitats.

The carbon values can be compared to values of extant *Potamochoerus porcus*, the African red river hog or bush pig, and *Hylochoerus meinertzhageni*, the African giant forest hog, if 1.5‰ is added to modern values. Harris and Cerling (2002) report values of –17.9 ± 3.5 for *Hylochoerus* and –9.4 ± 5.3 for *Potamochoerus*. *Propotamochoerus* values are slightly more positive than the most enriched forest hog values but fall within the range of values for bush pigs (see Figure 5.3). These results suggest that the Siwalik suid was not feeding on forest floor browse like the giant forest hog but rather had a varied diet that included fruit in wooded habitats. Furthermore, *Propotamochoerus* $\delta^{13}C$ values fall within the Cerling et al. (2003) category of hyperbrowser/frugivore for the S-level (see Figure 5.1), and most fall within that category for the PS-level, although one standard deviation from the mean does extend into the browser category (see Figure 5.2).

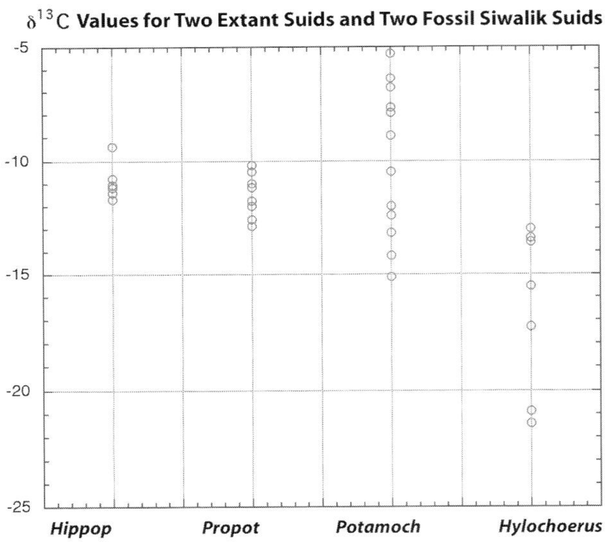

Figure 5.3 All observed $\delta^{13}C$ values of two Siwalik suids, *Hippopotamodon sivalense* and *Propotamochoerus hysudricus* are compared to values obtained from Harris and Cerling (2002) for two extant East African suids, *Potamochoerus porcus*, the African red river hog or bushpig (a frugivore) and *Hylochoerus meinertzhageni*, the giant forest hog (a folivore in closed habitat). 1.5‰ has been added to modern suid values to compensate for the shift in values of atmospheric CO_2 in the past 150 years due to burning of fossil fuels. The Siwalik suids are slightly more enriched compared to the forest hog but fall within the range of bushpigs, suggesting a range of habitats including woodland and a diet of browse and fruit.

Microwear

Forty-one specimens were analyzed, and most were characterized by numerous pits and coarse scratches, indicative of feeding on hard objects such as fruits or C_4 phytoliths. Isotopic analyses indicate a diet of exclusively C_3 vegetation prior to 8.0 Ma, and microwear posterior probabilities averaged across all specimens yielded the following results: *Potamochoerus* (.44) (fruit and roots); *Hylochoerus* (.36) (foliage), and; *Babyrousa* (.21) (fruit), thus favoring frugivorous suids. Of the 41 specimens, 24 were predominantly frugivorous (greater than 65 percent fruit), seven were predominantly folivorous (greater than 65 percent foliage), and 10 were mixed diets. These results suggest an omnivorous diet dominated by fruit and roots but with some periods of little fruit availability (microwear features turn over within a week). No change in diet is observed over time, but the two youngest specimens, dated to 7.1 and 7.0, are predominantly folivorous. Among the younger specimens, isotopic analyses are needed to determine whether heavy wear is still due to fruit rather than incorporation of some C_4 grass.

Thus, both isotopic and microwear analyses suggest *Propotamochoerus* had diet and habitat preferences similar to African bush pigs today. At the individual specimen level, for those sampled for both isotopes and microwear, three specimens yielding a predominantly frugivorous diet according to microwear (Y14502, Y4554, Y49515) are significantly more enriched in $\delta^{18}O$ values compared to three specimens with microwear indicating a greater proportion of browse (Y10224, Y11007, Y41368) (Mann-Whitney, corrected for ties: $Z = -1.96$; $p = .05$) (see Table 5.1). If dietary proportions did not change from the time of enamel formation to the time the individual died, then the enriched oxygen values may reflect feeding on enriched canopy fruit that has fallen to the forest floor. The depleted oxygen values of specimens suggesting more browse in the diet approach the oxygen range of giant forest hogs, suggesting browsing on closed forest floor vegetation. It should be noted, however, that only six specimens are represented, and specimens of other species thought to have fed on fruit based upon microwear evidence show a wide range of oxygen values.

Hippopotamodon sivalense: 10.2–7.2 Ma (10.2–7.1 Ma) Suinae

Isotopes

Hippopotamodon is also one of the most depleted S-level taxa for $\delta^{18}O$ values, with all but one of six specimens falling below the average value for the S-level (–5‰). The $\delta^{13}C$ values are average for that level. Among six S-level specimens, none show $\delta^{13}C$ values below –12‰, and one is less than –10‰, suggesting some foraging in open habitat but most foraging in woodlands. Only two specimens were sampled from the PS-level, and they do not differ from the S-level in carbon values. For both levels, most specimens fall within the hyperbrowser/frugivore category (see Figures 5.1 and 5.2). Some specimens are more enriched in $\delta^{13}C$ values for the S-level, however, and fall into the depleted portion of the browser category. Among the two PS-level specimens, one shows a very enriched (3.5‰) oxygen value, while the other is depleted (–4.1‰). The enriched tooth was sampled multiple times to confirm this enriched value. The results currently suggest that by 8.1 Ma, *Hippopotamodon* was exploiting a range of habitats including woodlands and open patches.

When compared to carbon values of modern African bush pigs and giant forest hogs, *Hippopotamodon* falls within the range of bush pigs. Thus *Hippopotamodon*, like *Propotamochoerus*, isotopically falls within the hyperbrowser/frugivore category and resembles a fruit-eating modern suid.

Microwear

Of 23 specimens analyzed, most were characterized by numerous pits and coarse scratches, indicative of feeding on hard objects. Posterior probabilities suggest a diet consisting primarily of fruit and some roots, in comparison to *Potamochoerus* (0.49), *Hylochoerus* (0.28), and *Babyrousa* (0.24). At the individual specimen level, 14 are primarily frugivorous, one is folivorous, and eight are mixed. These results suggest that both fruit and browse may have been a more consistent component of *Hippopotamodon* diet compared to *Propotamochoerus*, which indicated periods or geographic regions of little fruit availability. No change in diet is observed over time, but again, while isotopic analyses indicate all heavy wear features in specimens older than 8.0 Ma are due to fruit rather than C_4 phytoliths, isotopic analyses are needed to confirm a similar diet in younger specimens.

Thus, like *Propotamochoerus*, both isotope and microwear analyses indicate that *Hippopotamodon* was similar to African bush pigs in dietary and habitat preferences. But unlike *Propotamochoerus*, there is no evidence for seasonal change in diet.

Tetraconodon magnus: 10.0–9.3 Ma (11.2–9.3 Ma) Tetraconodontina

Isotopes

Tetraconodon is represented by one specimen in the S-level, and its $\delta^{13}C$ value of –12.8‰ is depleted relative to other taxa and indicative of feeding in forest. This places

Tetraconodon in the middle of the hyperbrowser/ frugivore range (see Figure 5.1). Its oxygen value is somewhat enriched relative to other taxa (–3.4‰ compared to an average of –5‰), suggesting less dependence on drinking water, feeding on fruit fallen from the uppermost canopy, or drinking in open habitat as well as forest. Note that this carbon-depleted forest species becomes extinct well before *Sivapithecus*.

Schizochoerus gandakasensis: 10.1–8.7 Ma (11.2–8.1 Ma) Doliochoerinae

Isotopes
Schizochoerus is represented by two S-level specimens. Both are very depleted in carbon, with values of –13‰ or lower, indicative of forest habitat. Oxygen values also fall below the average, with one as low as –9.2‰ (compared to average of –5‰). *Schizochoerus* is more depleted in both carbon and oxygen compared to all other taxa sampled, suggesting a preference for closed, wet habitat. It falls within the hyperbrowser/ frugivore category (see Figure 5.1). Note this forest species becomes extinct around the same time as *Sivapithecus*.

Microbunodon punjabiense: 12.3–9.2 Ma (14.0–8.6 Ma) Anthracotheriidae

Microbunodon, represented by three specimens, is also one of the most depleted taxa in the S-level. Two of the three specimens have $\delta^{13}C$ values below –12‰, indicative of forest habitat, and all oxygen values are depleted relative to the average, suggesting a habitat similar to that of *Schizochoerus*. Like *Schizochoerus*, *Microbunodon* also becomes extinct relatively early in the Late Miocene. Anthracotheres are considered ancestral to Hippopotamidae today. Hippos are aquatic by day and forage in open habitat at night, resulting in very depleted oxygen values but enriched carbon values in both modern and fossil species (Bocherens et al. 1996; Zazzo et al. 2000). *Microbunodon* shows the same affinity for aquatic habitats but not for foraging in open habitats.

Hemimeryx lg. sp.: 13.6–6.2 Ma (14.0–3.6 Ma) Anthracotheriidae

Isotopes
Hemimeryx is represented by one specimen in the S-level, and it is depleted in oxygen (–5.8‰) but very enriched in carbon (–8.5‰), thus joining *Hipparion* as the only S-level taxa completely within the browser category (see Figure 5.1). These results are compatible with the isotopic pattern of hippopotamus today, suggesting an aquatic habitat with feeding in open habitat.

Dorcabune nagrii: 10.4–8.5 Ma (11.2–8.4 Ma) Tragulidae

Isotopes
Five specimens from the S-level yield carbon results between –11.1 to –9.6‰, suggesting a preference for a range of habitats including woodland and some open patches. These values fall within both the hyperbrowser/frugivore category and the most depleted values of the browser category (see Figure 5.1). Oxygen values, spanning –8.2 to –1.6‰ could reflect foraging in a wide range of habitats, or they could reflect feeding on vegetation which includes fallen fruit from an enriched upper canopy.

Microwear
Ten specimens suggest a diet composed of at least 70 percent fruit on average. Sixty percent of specimens indicate diets of less than 35 percent browse, and none of the specimens have probabilities indicative of a browse component over 65 percent. These results suggest a diet as frugivorous as modern ruminants in rainforests today (Emmons 1983; Gautier-Hion et al. 1980).

Thus isotopic and microwear analyses indicate that this tragulid inhabited woodlands and fed predominantly on fruit.

Dorcatherium majus: 10.4–7.0 Ma (10.6–6.9 Ma) Tragulidae

Isotopes
D. majus is represented by five samples in the S-level, with carbon values spanning –12.7 to –10.5‰, indicative of woodland habitat with some forest. Most specimens fall within the hyperbrowser/frugivore category, with some

ranging into the most depleted values for the browser category (see Figure 5.1). Oxygen values range from –8.3 to +0.5‰, similar to *Dorcabune nagrii*, again suggesting foraging in a wide range of habitats or feeding on enriched canopy fruit lying on the forest floor.

Five PS-level samples yielded a significant enrichment in carbon values with respect to S-level specimens, with a range of –10.6 to –8.8‰, suggesting a shift to foraging in more open habitats. The mean $\delta^{13}C$ value shifts from the hyperbrowser category to the browser category (see Figure 5.2). Furthermore, the range of oxygen values decreases to values only between –3.8 to –1.7‰ (for a range of 2.1‰ compared to a range of 8.8‰ in the S-level), suggesting a decrease in habitat variety or a decrease in water dependence which resulted in loss of very depleted $\delta^{18}O$ values.

Microwear

D. majus is represented by 32 specimens characterized by numerous pits and coarse scratches. Posterior probabilities averaged across all specimens indicate a diet composed of at least 70 percent fruit, as in modern frugivorous rainforest ruminants. However, two specimens indicate a diet composed primarily of browse (over 65 percent), suggesting periods of fruit shortage. No change in microwear is observed over time.

Combined isotopic and microwear analyses suggest a diet composed primarily of fruit, but a shift from forest and woodland to more open habitat over time. Among three specimens sampled both isotopically and for microwear, the one specimen with microwear indicative of a significant browse proportion (47 percent) (Y49530) also shows the most enriched $\delta^{13}C$ value of –9.9‰, suggesting foraging in open habitat (see Table 5.1).

Dorcatherium sp. D: (10.3–7.2 Ma) (10.3–7.1 Ma)
Tragulidae

Microwear

Thirty-one specimens yield a microwear pattern indicative of frugivory, with fruit comprising at least 70 percent of the diet. Over 70 percent of specimens yield browse percentages less than 35 percent, but two specimens yield browse percentages over 65 percent, as in *D. majus*, suggesting periods of little fruit availability. Specimens present after the extinction of *Sivapithecus* show significant decreases in number of pits and scratches, suggesting a shift to feeding on less hard objects. The browse component also shifts from 26 percent on average to 34 percent, but this shift is not statistically significant. These results suggest a diet similar to tragulids today, with fruit availability decreasing over time.

Dorcatherium "Y270": 11.1–9.2 Ma (11.2–8.6 Ma)
Tragulidae

Microwear

This species is represented by only two specimens, but both indicate a diet of 70 percent fruit, similar to modern rainforest frugivorous ruminants.

Dorcatherium nagrii: 9.3–6.8 Ma (9.5–2.2 Ma)
Tragulidae

Microwear

Eleven specimens suggest a diet composed of both fruit and browse, with browse comprising a greater component (over 40 percent) of the diet than in modern frugivorous ruminants. Seventy-three percent of specimens had posterior probabilities suggesting a browse component greater than 35 percent, and 27 percent had diets composed of over 65 percent browse. *Dorcatherium nagrii* is similar in size to diminutive tragulids today and therefore must have fed on highly nutritious browse, such as young shoots and leaves. Specimens suggest an increase in the browse component over time, from 42 percent prior to the extinction of *Sivapithecus* (n = 7) to 61 percent afterwards (n = 3), but this shift is not statistically significant. However, when all tragulids species are combined, there is a significant increase in browse over time. Note that this tragulid is present long after many other frugivores have gone extinct.

Tragoceridus spp.: 11.2–6.2 Ma (11.2–2.2 Ma)
Boselaphini

Isotopes

Five S-level specimens cluster tightly in $\delta^{13}C$ values, between –11.7 and –10.7‰, suggesting an affinity for woodland. These values fall entirely within the hyperbrowser/frugivore category (see Figure 5.1). Oxygen values, on the other hand, span –6.7 to +0.7‰, suggesting a wide range of habitats, or feeding on enriched fallen canopy fruit in addition to other woodland vegetation.

Eight PS-level samples from six individuals yield a much greater range of carbon values, between –12.4 and –9.8‰, indicative of feeding in forest, woodland, and some open habitat. These specimens range from the hyperbrowser/frugivore category into the browser category, suggesting a shift in the diet to incorporate more browse in more open habitat (see Figure 5.2). Oxygen values again span –7 to +0.9‰.

M2s and M3s were sampled from the same jaw in four cases, two in the S-level and two in the PS-level. For each jaw, the M3 is enriched in both carbon and oxygen values relative to the M2. This pattern indicates feeding in more open, more evaporative habitats as the individuals developed and possibly seasonal preferences as well.

Microwear

Eight specimens were characterized by heavy wear features that were probably associated with fruit eating before 8.0 Ma, in accordance with the isotopic hyper-browser/frugivore results. Younger specimens, however, show evidence of grazing, especially at 6.4 and 6.3 Ma. Isotopic evidence is needed to corroborate this shift.

Dorcadoxa porrecticornis: 8.9–7.0 Ma (8.9–6.5 Ma) Reduncini

Isotopes

Seven PS-level samples from four individuals yielded carbon values between –11.9 and –9.3‰, suggesting a range of habitats from forest/woodland to open patches. These $\delta^{13}C$ values place *Dorcadoxa* within both the hyper-browser/frugivore and browser categories, with the mean falling on the threshold (see Figure 5.2). Oxygen values yield a greater range than any other taxon, spanning –8.0 to +1.6‰. This range probably reflects a range of habitats, as do the carbon values.

M2s and M3s were both sampled for three jaws. Like *Tragoceridus*, carbon values become more enriched from M2s to M3s, suggesting feeding in more open habitats as the individuals developed. Oxygen values, however, are always more enriched in M2s compared to M3s. The discrepancy in oxygen trends can be interpreted in several ways. First, it is possible that the two bovid taxa had different birthing seasons. Second, the reduncine *Dorcadoxa* is smaller than the boselaphine *Tragoceridus* and could have grown much faster. Alternatively, the $\delta^{18}O$ values could indicate that *Tragoceridus* was always an obligate drinker, while *Dorcadoxa* was only an obligate drinker during the dry season, thus resulting in the negative correlation between carbon and oxygen values.

Microwear

Five specimens are characterized by more fine scratches than coarse, suggesting a diet comprised of browse. Younger specimens yield evidence for incorporation of C_4 graze. When the two specimens older than 8.0 Ma are compared to the three dated to 7.3 Ma or younger, the browse proportion decreases from a posterior probability of 0.61 to 0.07. The C_4 grazer probability increases from 0.03 to 0.25. Thus microwear evidence suggests that *Dorcadoxa* was a mixed feeder on browse and C_4 graze, with C_4 graze comprising a greater proportion of the diet over time.

Combined the isotopic and microwear evidence suggest that the earliest *Dorcadoxa* were browsers in a range of habitats and that this species eventually incorporated C_4 graze in its diet. All modern reduncines are C_4 grazers. Isotopic analyses of younger samples are needed to confirm the transition to grazing in open habitats.

"2-Keel"

Microwear

Only two specimens were sampled, and they suggest a diet of mixed browse and C_4 graze.

Elachistoceras khauristanensis: 11.5–7.4 Ma (14.0–7.3 Ma) Boselaphini

Microwear

Five specimens were characterized by fine scratches and a low number of scratches, yielding a posterior probability of 0.75 for the browse category. Numerous pits and a probability of 0.2 in the mixed category yield some evidence of hard-object feeding, most likely fruit and seeds given that *Elachistoceras* was too small to rely heavily on C_4 graze. This bovid most likely fed on very high quality browse (young shoots and leaves) and soft fruit.

Indeterminate small bovids

Microwear

Four of these seven specimens are characterized by fine scratches and numerous pits, much like a chimpanzee. The remaining specimens differ by a low number of pits. These small bovids were most likely feeding on soft fruit and highly nutritious browse. The three specimens present after the extinction of *Sivapithecus* fed on a much higher percentage of browse compared to the three older specimens (84 percent compared to 33 percent), suggesting a dietary transition over time and a decrease in fruit availability.

Rhinoceros

Isotopes

Five PS-level specimens yield a wide range of carbon values, from –13.3 to –8.3‰, spanning the hyperbrowser/frugivore and browser categories (see Figure 5.2). This variability suggests exploitation of a range of habitats including forest and very open patches. These results are compatible with the highly omnivorous diet of modern rhinos, which feed on fruits, leaves, branches, shrubs, and grass (Nowak 1999; Kingdon 1997). If Siwalik rhinos were feeding on C_4 grass at 8.0 Ma, it was a very small proportion of the diet.

Oxygen values for these specimens are very depleted and show a tight clustering, unlike highly variable carbon values, spanning –4.3 to –3.2‰ (compared to an average of –2.7‰ for all PS-level specimens). Modern rhinos depend upon bodies of water for temperature control, digestion, and scent communication (Kingdon 1997). The depleted Siwalik oxygen results are compatible with water dependence, and the small range of values are compatible with dependence on large bodies of water that would have been buffered from seasonal changes in $\delta^{18}O$ values. With respect to water dependence, the Siwalik rhinos are similar to the Siwalik anthracotheres, suggesting an affinity for large watering holes.

Gomphotheres

Both S- and PS-level (n = 9) specimens yield a range of carbon and oxygen values which suggest exploitation of forests, woodlands, and some open habitat. S-level specimens all fall within the hyperbrowser/frugivore category, while some PS-level specimens range into the browser category (see Figures 5.1 and 5.2). Oxygen values are relatively depleted compared to other taxa, suggesting a greater dependence on water. Gomphotheres are proboscideans similar to elephants today, and the Siwalik results are compatible with very diverse modern elephant diets and habitats, which include fruit, browse, and graze in forests, woodlands, and semi-desert scrub (Nowak 1999). The Siwalik gomphotheres yield no evidence for feeding on C_4 graze, however. The depleted oxygen values suggest use of watering holes, as in modern elephants (Kingdon 1997).

***Deinotherium* spp.**

Isotopes

Deinotheres are also elephant-like proboscideans. The one specimen from the PS-level looks very much like Siwaliks giraffes in yielding highly enriched oxygen values and depleted carbon values (see Figure 5.2). These results suggest feeding in forest/woodland, with water obtained from very enriched upper canopy vegetation.

***Bramatherium megacephalum*: 11.3–7.1 Ma (10.3–7.1 Ma) Sivatheriinae**

Isotopes

The Siwalik *Bramatherium* is similar to other fossil and modern giraffes in exhibiting a wide range of carbon values and very enriched oxygen values (Quade et al. 1995b; Zazzo et al. 2000). For both the S- and PS-levels, carbon values range from less than –14‰, indicative of forest habitat, to almost –10‰, indicative of more open habitat. This range suggests feeding throughout forest canopy and understory, and it places most specimens within the hyperbrowser/frugivore category, with some specimens ranging into the most depleted values of the browser category (see Figures 5.1 and 5.2). Consistently enriched oxygen values suggest that *Bramatherium* was obtaining most of its water from upper canopy vegetation, consistent with other giraffes, both fossil and modern (Quade et al. 1995b; Zazzo et al. 2000). PS-level specimens are significantly more enriched in $\delta^{18}O$ values compared to S-level specimens by 2.2‰ on average. This shift is consistent with the 2.3‰ enrichment of all taxa on average between the S- and PS-levels, suggesting a shift in climate, most likely a decrease in rainfall.

Chalicotherium

Isotopes

Two S-level specimens yield carbon values that indicate a woodland habitat and a hyperbrowser/frugivore diet. Oxygen values are somewhat enriched compared to other taxa. Chalicotheres are large perissodactyls, and isotopic values are consistent with feeding in wooded habitat and with water intake from enriched upper story leaves.

Equids

Isotopes

Twenty-nine Siwalik equids sampled span a time range of 10.1 to 6.3 Ma. Two samples dated to 9.7 and 9.6 Ma yield $\delta^{13}C$ values below –12‰, indicative of forest habitat, but most samples indicate woodland and open habitat. The first indisputable evidence for C_4 graze appears at 8.7 Ma. After the extinction of *Sivapithecus*, most samples indicate a mixed C_3/C_4 diet, with the exception of a 7.8 Ma tooth with a relatively depleted carbon value of –9.8‰. S- and PS-level *Hipparion* specimens are more enriched in both carbon and oxygen compared to other taxa, with S-level specimens falling into the browser dietary category and PS-level specimens falling into both browser and mixed diet categories (see Figures 5.1 and 5.2).

Intra-tooth sampling of oxygen isotopes yielded precipitation curves that indicate a rainfall regime similar to monsoonal southern China today. Values become enriched over time, while the shape of the curve does not change, suggesting a decrease in rainfall but no change in seasonality over time. Intra-tooth carbon and oxygen values show positive, negative, or no correlations among different teeth and between different regions of the same tooth in some cases, suggesting no consistent seasonal incorporation of C_4 graze in the diet. Carbon values do suggest that once C_4 graze was incorporated into horse diets, it was a portion of the diet year-round, as was C_3 browse.

Finally, seasonal birthing is suggested by different tooth positions (premolars and molars) among the profiles, indicating most births occurred during the wet season. This pattern is consistent with birthing in Burchell's zebra today (Nowak 1999) and suggests a predictable seasonal climate.

Faunal Turnover

Microwear and isotopic analyses indicate that Late Miocene fauna included throughout the record many frugivores as well as open-habitat browsers and, in younger levels, mixed C_3/C_4 feeders. Species preferred habitats and diets correlate with times of extinction, with closed-habitat taxa and frugivores going extinct first and open-habitat taxa prevailing until later in the record (see Table 5.2). Species going extinct around the same time as *Sivapithecus* or earlier include *Schizochoerus* and *Microbunodon*, which were characterized by depleted carbon and oxygen values indicative of closed wet forest habitat, *Tetraconodon* characterized by depleted carbon values also indicative of forest, and two frugivorous tragulids, *Dorcabune nagrii* and *Dorcatherium* "Y270."

Another cluster of extinctions occurs around 7.4–7.1 Ma. This extinction event includes the omnivorous but predominantly frugivorous *Hippopotamodon*; a large frugivorous tragulid, *Dorcatherium* "Sp. D.;" the very small bovid *Elachistoceras*, which probably required year-round very high quality young shoots and leaves in addition to some fruit; and the giraffe *Bramatherium*, which yielded some of the most depleted carbon values, indicative of forest habitat, and oxygen values suggesting feeding from the upper canopy of forests.

More extinctions occur between 7.0 and 6.5 Ma, including *Propotamochoerus*, an omnivorous suid that appeared to have fed predominantly on fruit, although some individual specimens indicated a diet of browse; the large frugivorous tragulid *Dorcatherium majus*; the tragulid *Dorcatherium nagrii* for which microwear suggested a diet of fruit and browse, with browse being a much larger

Table 5.2

Time sequence of Late Miocene extinctions

Extinction coinciding with *Sivapithecus* extinction 8.5 (8.4) Ma or earlier	Extinction between 7.4-7.1 Ma	Extinction between 7.0-6.5 Ma	Extinction after 6.5 Ma
Tetraconodon 9.3 Ma	*Hippopotamodon* 7.2 (7.1) Ma	*Propotamochoerus* 6.8 (6.5) Ma	*Hemimeryx* 6.2 (3.6) Ma
Schizochoerus 8.7 (8.1) Ma	*Dorcatherium* "Sp. D." 7.2 (7.1) Ma	*D. majus* 7.0 (6.9) Ma	*Tragoceridus* 6.2 (2.2) Ma
Microbunodon 9.2 (8.6) Ma	*Elachistoceras* 7.4 (7.3) Ma	*Dorcatherium nagrii* 6.8 (2.2) Ma	
Dorcatherium "Y270" 9.2 (8.6) Ma	*Bramatherium* 7.1 Ma	*Dorcadoxa* 7.0 (6.5) Ma	
Dorcabune nagrii 8.5 (8.4) Ma			

The time sequence of Late Miocene extinctions of taxa studied in these isotopic and microwear investigations suggests a response to loss of forest, with *Sivapithecus*, closed wet habitat species, and two tragulids going extinct first. Subsequent extinctions include more omnivorous but still predominantly frugivorous suids, a very small bovid, and more tragulids, with the most folivorous tragulid being one of the last to go extinct. Last occurrence of fossil specimens is followed by statistically estimated last appearances in parentheses, from Barry et al. (2002).

component of the diet than in modern tragulid diets; and *Dorcadoxa*, a bovid that inhabited woodland and open regions and fed on both browse and C_4 graze, with the C_4 component increasing over time. Among the taxa analyzed in these studies, *Hemimeryx* and *Tragoceridus* do not become extinct until after 6.5 Ma. *Hemimeryx* is isotopically similar to hippopotamuses today, and *Tragoceridus* specimens most likely represent multiple species, with later species feeding on both browse and C_4 graze.

Thus during the time interval under investigation, isotopic and microwear analyses suggest a decrease in annual rainfall over time, accompanied by gradual loss of forest and an increase in open habitat, including patches of C_4 graze. These analyses do not suggest complete replacement of forest until some time after 7.0 Ma. Timing of extinctions, in addition to isotopic evidence for a changing vegetation mosaic, suggest forest fragmentation well before complete forest replacement. In modern disturbed ecosystems, large frugivores are the first species endangered (Johns and Skorupa 1987). Frugivores in general suffer, as demonstrated by a study of logged areas of Kibale Forest. In these regions, biomass densities of frugivorous Kibale primates averaged 59 percent lower than in undisturbed forest, compared to a decrease of 39 percent for folivorous primates (Skorupa 1986). Additionally, many frugivores are particularly sensitive to decreases in fig tree densities, for these trees provide fallback foods during periods of preferred fruit shortages (Johns and Skorupa 1987). Effects of selective logging were also studied in rainforests of northern Borneo, where tragulids appear to be affected as adversely as frugivorous forest-dwelling primates (Heydon and Bulloh 1997). In the tragulid investigation, two species studied were both less common in logged forest, with the species most dependent on large strangling fig trees exhibiting the greatest decline.

Furthermore, fragmentation of habitats adversely affects animals in general, for small isolated habitats support smaller populations, and the smaller the population, the greater the susceptibility to extinction due to demographic and environmental stochasticity (Newmark 1985, 1987, 1996). Also, species with slower intrinsic rates of growth, such as large-bodied apes, are more susceptible to extinctions than animals which reproduce more quickly, suggesting another reason why *Sivapithecus* may have been more vulnerable to initial habitat change than other Late Miocene species (Diamond 1984).

The Siwalik Late Miocene sequence of extinctions suggest that frugivorous and closed-habitat species with different thresholds for forest size and fruit abundance requirements go extinct at different times, in accordance with the modern species responses to forest disturbance (see Figure 5.4). Among the first species to go extinct are the closed, wet habitat species; *Sivapithecus*, a large-bodied frugivorous ape; and two tragulids. Subsequent extinctions include more omnivorous but still predominantly frugivorous suids, and more tragulids, with the most folivorous tragulid of this investigation being one of the last to go extinct. This gradient of extinctions is most likely a response to forest fragmentation, with species responding at different times due to different susceptibilities.

Localities

Several fossil localities yielded striking microwear or isotopic results, and most of these localities were sampled using both techniques. The S-level locality Y182 was represented by multiple *Sivapithecus* and *Propotamochoerus* specimens that were characterized by very heavy wear including very high pit counts. These results suggest that these specimens were feeding on fruits with hard exocarps or seeds. Isotopic analyses of teeth from this locality yield $\delta^{13}C$ values indicative of forest and woodland, and *Schizochoerus*, the indicator species of closed, wet forest, is present. Specimens from the S-level locality Y269, including *Schizochoerus*, are depleted in both carbon and oxygen, indicating the presence of closed, wet forest. *Propotamochoerus* specimens analyzed for microwear for locality Y269 include two specimens that appear highly frugivorous and two that suggest a greater proportion of browse in addition to fruit. At both Y182 and Y269, *Propotamochoerus* specimens yield the most enriched carbon values, suggesting this suid fed in woodland habitat associated with the forests. Locality Y312 isotopic results indicate that this locality was unusually open for the S-level. Only one *Tragoceridus* specimen was sampled for microwear from this locality, and it suggests a diet of both browse and fruit. *Tragoceridus* also yields the most depleted carbon values for this locality, suggesting that it preferred the woodlands associated with the more open regions of Y312. Locality Y917, dated to 7.8 Ma, yielded a horse specimen unusually depleted in both carbon and oxygen for this time, indicative of a pure C_3 diet despite evidence for C_4 graze among older horse specimens. Microwear analysis of a *Hippopotamodon* specimen from this locality indicates a diet with a very strong fruit component, also suggesting less open habitat. Finally, locality Y258, dated to 10.1 Ma, yielded an isotopically very enriched horse specimen for this time level, corroborating lithological evidence that this locality was unusually open, but no specimens from this locality were sampled for microwear.

Some localities therefore yield consistent isotopic values or microwear patterns, and for some of these, isotopic and microwear results complement one another. This suggests that these localities were temporally and spatially more similar to specific points in time and specific geographic regions rather than averages of surrounding habitats sampled over many thousands of years.

Correlation of Results with Previous Paleosol Studies

Previous isotopic analyses of Siwalik paleosols yielded $\delta^{13}C$ and $\delta^{18}O$ values from which vegetation mosaic patterns, including the appearance of C_4 grasses, and climatic changes were inferred (Quade and Cerling 1995). These results provide a comparison to the isotopic analyses of mammalian teeth in this investigation (see Figure 5.4). With respect to the C_3/C_4 transition, paleosol samples first detect isotopic evidence for C_4 grasses on the floodplain at 8.1 Ma. The first indisputable dietary evidence for C_4 graze in mammalian teeth occurs at 8.7 Ma, but somewhat enriched carbon values as early as 9.3 Ma

Age (Ma)	Climate Change	Vegetation Change	Faunal Change
6.0			
6.1			
6.2			
6.3			
6.4			
6.5			Frugivorous suid and tragulid extinctions
6.6			
6.7			
6.8	Decrease in annual rainfall	**First C4 grasslands (paleosols)**	
6.9			
7.0		**Last C3 dominated vegetation (paleosols)**	
7.1			Frugivorous suid, tragulid, small bovid, and forest habitat giraffe extinctions
7.2			
7.3			
7.4		**First C4 dominated vegetation (paleosols)**	
7.5	**Decrease in Nepal rainfall inferred from bivalves**		
7.6		First C4 in bovid diets	
7.7	**First highly enriched precipitation curve (equid)**		
7.8			
7.9			
8.0	**Last paleosols with depleted oxygen**	**Permanent patches of C4 grasses inferred from C4 evidence in most equid diets year-round**	
8.1			
8.2		Forest fragmentation	
8.3	**Last depleted precipitation curve from equid tooth**		
8.4			
8.5			
8.6			
8.7		**Small patches of C4 grasses inferred from first indisputable dietary evidence for C4 (equid)**	Closed, wet habitat taxa, frugivorous tragulids, and *Sivapithecus* extinctions
8.8			
8.9			
9.0			
9.1			
9.2	**Start of paleosol oxygen enrichment**		
9.3	**First enriched precipitation curve from equid tooth**		
9.4	**Beginning of summer Arabian upwellings**		
9.5			

Figure 5.4 Synthesis of Late Miocene changes in climate, vegetation, and fauna in the Siwaliks.

suggest C_4 grasses may have been a very small component of some diets prior to 8.7 Ma, in agreement with tooth analyses conducted by Morgan (1994). Evidence from the 8.7 Ma equid tooth, sampled multiple times along the length of the tooth, indicates that both C_3 and C_4 plants contributed to the diet year-round. However, equid teeth sampled similarly at 8.5, 8.4, and 8.3 Ma indicate periods of the year in which no C_4 component is present. It is not until 8.0 Ma and later that intratooth sampling of every equid tooth detects a C_4 graze component year-round, that is, in every sample taken from the tooth. Thus, both paleosol and equid enamel isotopic samples indicate the first permanent C_4 patches around 8.0 Ma or slightly earlier. Even at 8.0 Ma, however, no bovids sampled were feeding on any C_4 graze, suggesting that C_4 grasses did not become extensive until some time after 8.0 Ma. Paleosols indicate the first C_4 dominated habitats around 7.4 Ma, with the last C_3 dominated habitats at 7.0 Ma and the first C_4 grasslands at 6.8 Ma (Quade and Cerling 1995, corrected to Cande and Kent 1995; summarized in Barry et al. 2002). Microwear analyses detect a C_4 graze component for the bovid *Tragoceridus* at 6.4 Ma, but no specimens were analyzed between 8.1 and 6.4 Ma. Likewise, microwear detects a C_4 graze component in the bovid *Dorcadoxa* at 7.3 Ma, but no specimens between 8.0 and 7.3 Ma were analyzed. Thus bovids began to incorporate C_4 graze in their diets some time between 8.0 and 7.3 Ma, and paleosols suggest the first C_4 dominated vegetation mosaics at 7.4 Ma, coinciding with microwear evidence.

Paleosol oxygen isotopic analyses provide evidence of climatic change, with enrichment of $\delta^{18}O$ values over time. This shift is sporadic, with both depleted and enriched values present through much of the record, similar to the fluctuating pattern of enrichment detected in horse tooth profiles between 9.3 and 8.4 Ma. These fluctuations probably reflect a highly variable vegetation mosaic, with horses drinking in both closed and open habitats, as well as potential fluctuations in annual rainfall amounts. The oxygen transition begins in the paleosols at 9.2 Ma, with the first $\delta^{18}O$ values that are more positive than –8‰ (Quade and Cerling 1995, corrected to Cande and Kent 1995; summarized in Barry et al. 2002). Among the eleven horse tooth profiles that yielded seasonal curves apparently unaffected by tooth histology or buffered water sources, the first shift from depleted curves to moderately enriched ones occurs at 9.3 Ma, in accordance with the 9.2 Ma initial transition observed in the paleosols. Within the paleosol record, almost all samples are greatly enriched after 8.0 Ma compared to older samples. Within the horse tooth profiles, the last depleted curve occurs at 8.3 Ma, and the first highly enriched curve occurs at 7.7 Ma, thus surrounding the enrichment shift observed in the paleosol record. A 3.7‰ enrichment is observed in Late Miocene paleosols. This shift is less than the 5.6‰ shift observed in the horse tooth profiles, but the paleosol shift is confounded by annual seasonal differences, whereas the profiles allow comparisons of particular seasons, such as shifts in $\delta^{18}O$ values of the driest or wettest months over time.

With respect to the vegetation mosaic, paleosol carbon isotopic values directly represent vegetation growing above the soil, averaged over 10^2–10^4 years. Oxygen values are determined by the $\delta^{18}O$ value of meteoric water, the degree to which this water undergoes evaporative effects, and by temperature (Quade and Cerling 1995). Tooth enamel values, on the other hand, represent only the particular diet of the animal at a single point in time. These two approaches thus provide two views of the paleovegetation at the time, one a broad-scale record of vegetation across the paleoecosystem, the other a fine-scale record as seen by the herbivorous mammals that could move about the landscape and feed from specific patches of vegetation.

A total of 96 fossil teeth were sampled between 9.3–8.0 Ma, while only 12 paleosol carbonate samples have been analyzed between the ages of 9.4–8.0 Ma. Carbon values of tooth enamel for the S-level (9.3–9.2 Ma) ranged from –14.6 to –8.1‰. Paleosol carbonates between 9.4–9.2 Ma range between –12.6 to –10‰. The enrichment factor for $\delta^{13}C$ values between vegetation and paleosol carbonate is about 15‰, compared to 14‰ for enamel. A subtraction of 1‰ from paleosol values is therefore needed to compare them to enamel, shifting their range to –13.6 to –11‰. Both methods therefore detect forest and woodland, but only enamel analyses detected open habitat. This discrepancy is likely due to the very small paleosol sample size, but could suggest that open habitats were not an extensive part of the ecosystem. At the PS-level (8.1–8.0 Ma), enamel yielded a range of

$\delta^{13}C$ values between –14.2 and –5.3‰. Paleosol samples fall between –11.6 to –8.9‰ (or –12.6 to –9.9‰ to compare to enamel values). Again, enamel analyses detected more forest and more open habitat than did paleosol analyses, and only enamel analyses detected the presence of C_4 grasses. This discrepancy is also likely due to the small paleosol sample size, but it could also suggest that both forests and C_4 grasses were not extensive at this time level.

S-level enamel samples ranged in $\delta^{18}O$ values between –9.2 and +0.7‰, while paleosol values ranged between –9.9 and –5.4‰. Many factors could have led to enrichment of oxygen values in enamel with respect to meteoric water values, including enrichment due to metabolic processes, feeding on enriched vegetation, and drinking from bodies of water undergoing evaporation. Only evaporation and temperature affect paleosol carbonate oxygen values. Therefore, the best comparison between results from these two methods lies in the most depleted samples. These depleted values, –9.2‰ for enamel and –9.9‰ for paleosols, are very similar, suggesting that these values provide the best approximation of unevaporated soil water for this time interval. Enamel and paleosol depleted oxygen values are again very similar at the PS-level (–8.0 and –8.9‰ respectively), also suggesting the best approximation of unevaporated soil water. The difference between S- and PS-level depleted samples suggests an enrichment of at least 1‰ for soil water over time, probably more so given groundwater is buffered, and is likely an indication of a shift in the rainfall regime with a decrease in annual rainfall.

Finally, morphology of paleosols corroborates evidence for both a decrease in rainfall throughout the Siwaliks and a seasonal rainfall regime throughout the Late Miocene. Clay minerals from Middle Miocene paleosols older than 14 Ma are dominated by kaolinite, which is present in Indian soils in wet climates (greater than 1,000 mm rainfall per year). Kaolinite becomes less dominant over time, suggesting a decrease in rainfall over time (Quade and Cerling 1995). Climatic modeling predicts a decrease in rainfall in this geographic region with the increasing rise of the Tibetan Plateau (Prell and Kutzbach 1992). Alternatively, the supplemental winter rains in Pakistan today which result from westerlies carrying moisture from the Mediterranean may have been more substantial in the Late Miocene, making the "dry" winter months wetter than today. Furthermore, Quade and Cerling (1995) suggest that the presence of soil carbonates from 12 to almost 2 Ma establishes an upper boundary of 1,250 mm annual rainfall throughout the Siwaliks, given that carbonates are not generally present in soils in North America and Europe where rainfall exceeds this amount. Finally, in the Chinji Formation (14.2–12 Ma), soil morphology indicates periods of intense oxidation (Behrensmeyer et al. 1995). Behrensmeyer et al. (1995) suggest that a monsoonal rainfall regime with prolonged dry seasons would have promoted seasonal fluctuations in the water tables and resulted in the observed oxidation of organic materials.

Figure 5.4 summarizes changes in climate, vegetation, and fauna over time in the Late Miocene. Microwear and isotopic evidence from this research, in conjunction with isotopic paleosol evidence, suggests that the Late Miocene was characterized by a seasonal rainfall regime which was likely monsoonal, with a decrease in rainfall over time accompanied by forest fragmentation and replacement by open habitat, including C_4 grasses. A decrease in rainfall began around 9.3 Ma and took place in phases, with a second significant decrease in rainfall between 8.0 and 7.7 Ma. C_4 grasses appeared on the floodplain by at least 8.7 Ma, but patches of C_4 grass did not become extensive until after 8.0 Ma. As forests were replaced by more open habitat, closed-habitat species and frugivorous species became extinct. Frugivore extinctions also occurred in phases according to thresholds for different species, with the first phase of most susceptible species including the large-bodied frugivorous hominoid *Sivapithecus*.

Sivapithecus Habitat

Microwear analyses suggest that *Sivapithecus* was similar to modern frugivorous great apes and that many contemporaneous species were as frugivorous as their modern counterparts living in rainforests today. However, isotopic evidence suggests that even as early as the S-level, when *Sivapithecus* was present, the habitat was a vegetation mosaic in which there was forest, but not a completely closed canopy rainforest. Furthermore, the annual rainfall regime inferred from equid intra-tooth $\delta^{18}O$ curves is more seasonal, with a greater number of dry months (five-to-six), than any regime that almost all extant great apes experience today (four dry months) or any regime that supports evergreen rainforest today (no more than four dry months). These results suggest that *Sivapithecus* habitat was similar to monsoonal forests in southern China today in terms of vegetation mosaic and rainfall regime.

If the spatio-temporal availability of fruit in Siwalik monsoon forests were similar to availability in modern monsoon forests, then *Sivapithecus* must have differed from modern apes in habitat requirements. In this scenario, either *Sivapithecus* required less fruit or relied more heavily on fallback foods than modern great apes, or it was better able to travel greater distances in search of fruit. Microwear analyses suggest that *Sivapithecus* was very similar to orangutans and chimpanzees in terms of diet, and both of these species rely heavily on fruit. *Sivapithecus* is also smaller than gorillas and is therefore less likely to have relied on less nutritious foods than fruit. Furthermore, *Sivapithecus* had a life history pattern similar to modern great apes (Kelley 1997). Given its protracted reproductive schedule and therefore inferred preference for high quality foods, in combination with the microwear results, it is unlikely that *Sivapithecus* required less fruit than chimpanzees or orangutans. Furthermore, *Sivapithecus* postcrania indicate no locomotor specializations, such as brachiation, knuckle-walking, or even much suspensory behavior, that would have allowed it to be more efficient or even as efficient as modern apes in ranging in search of fruit.

The only great apes that inhabit or may have once inhabited environments as seasonal as the one suggested for the Late Miocene of the Siwaliks are a few populations of chimpanzees and possibly some extinct populations of orangutans. Chimpanzee populations in Mt. Asserik, Senegal, experience a dry season of at least five consecutive months, and mean annual rainfall averages 824–1,220 mm compared to averages of 1,500 mm at other chimpanzee sites (McGrew et al. 1981). Mt. Asserik population densities are extremely low, with a home range size estimated at 278–333 km^2, compared to no more than 40 km^2 for chimpanzees at Gombe and Kasoge (Baldwin 1982 et al.). These chimpanzees must travel long distances to reach dispersed fruit during the wet season, and they are confined to gallery forest during the dry season (Baldwin 1982 et al.). While no *Sivapithecus* postcranial features preclude terrestrial locomotion, *Sivapithecus* shows no specializations for extensive terrestrial trekking (Kelley and Pilbeam 1986), and it therefore probably did not travel the distances Mt. Asserik chimpanzees do today.

Fossil evidence indicates that orangutans inhabited a much larger geographic range in the Pleistocene compared to present day, extending into the interior of China. These populations may have experienced more seasonal climatic conditions than apes do today, but even Pleistocene orangutan populations appear to have tracked tropical forests, with their geographic range expanding and contracting throughout the Pleistocene in response to latitudinal movements of vegetation belts (Jablonski et al. 2000). Modern orangutans experience greater temporal fluctuations in fruit availability than other apes due to the masting phenomenon of southeast Asian forests, but they are able to cope by storing large amounts of energy as fat and therefore take advantage of fruit superabundance (Knott 1998).

It is possible that *Sivapithecus* shared this fat storage adaptive response, and masting may have characterized Siwalik forests. Dipterocarpaceae, the dominant family of Bornean rainforest trees, display the mast-fruiting reproductive strategy, and pollen evidence suggests that dipterocarps were present in the Nepal Miocene Siwaliks. However, evidence suggests that masting years in southeast Asia generally occur during El Nino-Southern Oscillation (ENSO) events (Curran et al. 1999), with this correlation weakening toward the western edge of Malaysia (Wich and Van Schaik 2000). If the phenomenon of unpredictable fruiting years is a response to an unpredictable climate, the ENSO events are a Pacific Ocean phenomenon not likely to have had an effect in Pakistan. Furthermore, the monsoonal system suggested by precipitation reconstructions in this thesis might have yielded a more predictable climate regime than that of regions affected by ENSO events. However, the monsoonal system today does show some variance, with dry years and years of floods, and it is possible that this system may have been more variable during its establishment in the Miocene.

Thus, *Sivapithecus* exhibits no specializations for efficient locomotion to reach highly dispersed fruits, and there is no current evidence to suggest that it inhabited a masting environment or had fat storage capabilities.

There is therefore no current evidence to suggest that *Sivapithecus* might have differed in habitat and dietary requirements from the range of variation currently exhibited by extant great apes.

An alternative hypothesis is that the Miocene forests differed in the spatio-temporal availability of fruits compared to monsoonal forests today, thereby mimicking modern ape rainforests. Four scenarios might have established this difference. First, while the type of forest today is most dependent on length of the dry season (Richards 1996), and the dry season suggested for the Siwaliks would result in a monsoon forest rather than a rainforest, amount of annual rainfall also plays a role, and one hypothesis is that amount of rainfall during the Late Miocene was greater than that experienced by forests today, thereby compensating for length of the dry season. Paleosol evidence indicating less than 1,250 mm mean annual precipitation suggests that this was not the case (Quade and Cerling 1995), and vegetation mosaic reconstructions in this thesis suggest little to no closed canopy rainforests.

A second scenario is that there was a greater diversity of fruiting species in Miocene forests. The few populations of gibbons that inhabit monsoonal forests appear to prefer environments with high tree diversity because these habitats buffer the effects of seasonality by enabling animal-dispersed fruit to be available throughout the year (Chivers 1984). A diversity of fruiting species with different microhabitat requirements could also have decreased the spatial patchiness of fruit. In particular, a greater diversity of asynchronous fruiting species such as *Ficus* would have increased the temporal availability of fruit. The significance of these fallback fruits is illustrated by the sensitivity of frugivorous primates and tragulids to *Ficus* abundance in disturbed forests (Heydon and Bulloh 1997; Johns and Skorupa 1987).

A third scenario is that like modern apes, *Sivapithecus* would have inhabited gallery forests, or forests associated with rivers, and these gallery forests may have been more extensive in the Miocene Siwaliks than in any regions inhabited by great apes today. Perhaps Siwalik forests resembled those of modern Amazon or Okavango river basins. These regions are characterized by very fertile soils that can support greater spatial and temporal availability of fruits and thus might support frugivorous fauna year-round. Furthermore, many of these regions, such as Manu National Park in Peru, are characterized by heavy tree fall, low canopy, and an understory of vines (Terborgh 1990). This vegetation mosaic is consistent with the mosaic inferred from the isotopic analyses in Chapter Three, and lianas might provide a pathway for *Sivapithecus* to travel throughout the canopy.

A fourth scenario, suggested by Janis et al. (2000, 2002) to explain high diversity among Middle Miocene browsing species in North America, is that plant productivity may have been higher in the Miocene if atmospheric CO_2 (pCO_2) levels were higher. A higher plant productivity would have increased availability of young, highly nutritious leaves, thereby supporting the more folivorous small bovid and tragulid diets inferred from microwear, and perhaps increasing fallback food possibilities for frugivores. This argument relies on two assumptions: 1) plant productivity increases with increasing pCO_2, and; 2) pCO_2 levels were higher in the Miocene. If this scenario were correct, its effects should have been global rather than local.

Laboratory studies have shown that elevated CO_2 levels promote increased productivity levels in plants because higher CO_2 allows for greater rates of photosynthesis and vegetation growth, as well as increasing water use efficiency (Bazzaz 1990, Ehleringer et al. 1991; Ehleringer et al. 1997). CO_2 enrichment of non-laboratory ecosystems using free-air CO_2 enrichment technology (FACE) (a ring of gas-delivery systems surrounding vegetation plots) in both North Carolina pine forests and the Mojave desert also yielded increased plant productivity. Furthermore, in the Mojave desert, higher CO_2 also resulted in successful invasion of exotic species, thus increasing species diversity (DeLucia et al. 1999; Smith et al. 2000). An increase in water use efficiency is particularly intriguing because this increase is almost directly proportional to the level of CO_2 for a given temperature and humidity regime (Farquhar 1997). This means that doubling pCO_2 would have an effect similar to doubling rainfall for the plant. A higher level of pCO_2 in the Miocene may have allowed for modern rainforest plant species to have lived under monsoon forest rainfall conditions.

Finally, it is unclear how higher CO_2 levels may have affected phenology because most experimental studies were terminated before reproduction. In the few studies on flowering time, elevated CO_2 levels sometimes decreased time to flowering, sometimes increased it, and sometimes had no effect based upon the species (Bazzaz 1990). In a study of elevated CO_2 effects on *Datura stramonium*, fruit weight increased, seed size was unaffected, and plants produced fruits with thicker walls (Bazzaz 1990). These results must be treated with caution, for they only represent the response of one species, but they suggest that higher CO_2 levels, in addition to supporting higher plant productivity, may support greater fruit productivity for some species as well. An increase in fruit walls may also have ramifications for frugivore tooth morphology and microwear, leading to very coarse features, as observed in many Siwalik taxa.

Elevated Miocene pCO_2 levels, with a decrease in pCO_2 to present day levels during the Late Miocene, were first proposed as the mechanism by which plants with a C_4 photosynthetic pathway suddenly became globally abundant around 8–7 Ma (Cerling et al. 1993, 1997b; Ehleringer et al. 1991, 1997). Conditions determining competitive advantage between C_3 and C_4 plants are temperature and pCO_2, with C_4 plants favored under low pCO_2 conditions when accompanied by warm growing seasons. When pCO_2 levels are above 500 parts per million by volume (p.p.m.v.), the C_3 photosynthetic pathway is favored under all climatic conditions. Current pCO_2 levels are about 350 p.p.m.v., thus giving C_4 plants a competitive advantage in warm, dry climates, and it has been proposed that it was a crossover of the pCO_2 threshold to values under 500 p.p.m.v. in the Late Miocene that allowed C_4 plants to become abundant (Ehleringer et al. 1997). A shift in pCO_2 would explain why abundant C_4 biomass suddenly appeared in Africa, Asia, and North America between 8–6 Ma, evolving multiple times within eight of 16 superorders in Dicotyledonae and in one of five superorders of Monocotyledonae.

It has been proposed that a Late Miocene decrease in pCO_2 was caused by an increase in chemical weathering rates due to uplift of the Himalayas and the Tibetan plateau, which caused a drawdown of atmospheric CO_2 (Filippelli 1997; Raymo and Ruddiman 1992). Several lines of evidence indicate an intensification of the Asian monsoon and Tibetan uplift around 9–7 Ma, including Arabian Sea summer upwellings (Kroon et al. 1991), peaks in Indian Ocean sediment flux rates which were of Himalayan-Tibetan origin (Rea 1992), and increases in dust accumulation in China (Guo et al. 2002). What has proven difficult to measure is the actual pCO_2 levels in the Miocene. Attempts at Miocene pCO_2 estimates have included evidence from biomarker production by planktonic foraminifera (Pagani et al. 1999), boron-isotope ratios of planktonic foraminifera to estimate oceanic pH levels (which respond to changes in pCO_2) (Pearson and Palmer 2000), and leaf stomatal indices in fossilized plants. Each of these studies concludes that pCO_2 levels had already dropped to modern pre-industrial levels before the Late Miocene. However, there are issues with each study that make interpretation of their results problematic.

Studies of planktonic foraminifera assume that ocean pH, calcium carbonate levels, and CO_2 levels will respond rapidly to changes in pCO_2. However, Anderson and Archer (2002) have found that at the timescale of thousands of years, ocean pH response to changes in pCO_2 is highly variable. In their study of ocean pH and carbonate ion concentrations during a Pleistocene glacial/interglacial shift in pCO_2, they found that the expected corresponding pH shift occurred in the Atlantic Ocean, but there was no shift observed in the Pacific or Indian Oceans. Their results suggest that ocean response to pCO_2 is driven at least in part by differences in ocean currents. The amount of carbon stored in the atmosphere, land, and surface ocean is less that 10 percent of the inorganic carbon stored in the deep ocean (Sigman and Boyle 2000). Deep water turnover, thereby exposing it at the ocean surface, occurs only every 1,000 years, meaning that changes in pCO_2 that are not driven by ocean processes will be diluted by the large carbon reservoir of the deep ocean, at least on the timescale of thousands of years (Sigman and Boyle 2000). It is possible that the Atlantic Ocean responded to pCO_2 shifts when other oceans did not because it is one of the few places where deep water forms, creating a surface to deep water conveyor belt that can change in circulation in only 300 years or less (Boyle 1990). Finally, ocean pH levels are determined not only by biogenic precipitation of carbonate, which removes alkalinity, but also by weathering of emergent carbonates on land, which adds alkalinity (Sigman and Boyle 2000), suggesting the weathering rates of the Himalayas might be a factor to consider in making inferences about ocean pH for the Late Miocene.

Estimates of pCO_2 from fossilized stomata are based upon the relationship between leaf stomatal abundance and growing season CO_2 levels (Retallack 2001; Royer et al. 2001). These studies determined that stomatal indices in fossil plants related to modern Ginko correspond to present-day pCO_2 levels in the Miocene. These studies are hindered by the small range of species analyzed, for stomatal response to CO_2 varies greatly among species and is influenced by environmental factors other than pCO_2, including soil moisture and light levels (Bazzaz 1990). Furthermore, the mechanism by which CO_2 controls stomatal activities is unknown.

Thus while high pCO_2 levels in the Miocene followed by a decrease in pCO_2 levels would provide a mechanism by which C_4 plants could have suddenly become abundant globally, pre-decrease levels could support North American vegetation capable of nourishing a diversity of browsers unparalleled today, and the Siwalik monsoon forests could have supported a range of frugivorous species including a hominoid, there is no current evidence to support this scenario. While attempts at measuring Late Miocene pCO_2 levels indicate that they had already approached modern levels, these inferences have complications, and further research is needed to determine more confidently pCO_2 in the Late Miocene. If a decrease in pCO_2 levels was not responsible for the vegetation and faunal patterns of the Late Miocene, a new mechanism is needed, especially to explain the sudden global expansion of C_4 grasses. Alternative explanations include changes in global seasonal precipitation regimes to more warm-season precipitation or increased aridity. Evidence from the Siwaliks does not support a seasonal shift at the time of C_4 expansion, but it does support a decrease in rainfall.

Conclusions

Isotopic evidence from a wide range of large mammalian species indicates that the Siwaliks were characterized by a vegetation mosaic of forest, woodland, and open habitats as early as 9.3 Ma. There is no indication of completely closed canopy rainforest. By 8.1 Ma, much of the forest was replaced, but not completely eliminated, by more open habitat, including patches of C_4 grass. These patches of C_4 grass did not become extensive until some time after 8.0 Ma.

This vegetation mosaic supported fauna that exploited a range of habitats. Microwear analyses indicate that many species, including suids, tragulids, and bovids, were as highly dependent on fruit as their modern rainforest counterparts. Likewise, microwear results suggest that *Sivapithecus* was as dependent on fruit as modern frugivorous great apes. Most of these species go extinct over the time interval sampled in this thesis, with timing of extinction correlated to closed habitat and fruit requirements. Those species that appeared to have been most dependent on forest or fruit, including *Sivapithecus*, are the first to go extinct in the Late Miocene. This pattern, combined with vegetation mosaic reconstructions, suggests loss of forest over time, which then led to extinction of frugivores, including *Sivapithecus*.

Isotopic reconstructions of the precipitation regime from 10.0 to 6.3 Ma are very similar to the regimes experiences in monsoonal southern China today. These reconstructions suggest a dry season of five-to-six months, with an intense rainy season. It is therefore likely that the modern Asian monsoon had begun development as early as 10 Ma. An enrichment of oxygen value over time suggests a decrease in rainfall that was substantial enough to result in a change in forest type over time.

Combined isotopic evidence of the vegetation mosaic and seasonal precipitation regime suggests a habitat most similar to monsoonal forests today, with a shift from wet monsoonal to dry monsoonal forests from 10 Ma to 6.3 Ma. Monsoonal forests do not support great apes today. There is no current evidence that *Sivapithecus* might have differed in habitat and dietary requirements from those of extant great apes. Therefore, it is possible that Miocene forests differed in the spatio-temporal availability of fruits compared to modern monsoonal forests, thereby mimicking modern ape forests that support a wide range of frugivorous species. Miocene monsoon forests may have consisted of a greater diversity of fruiting species than modern forests, thereby supporting frugivores, or they may have been associated with rivers which provide fertile soil that can support greater fruit productivity. Alternatively, it has been proposed that the C_3/C_4 transition was a result of higher atmospheric CO_2 levels in the Late Miocene that fell to modern levels around 8 Ma (Cerling et al. 1993, 1997b; Ehleringer et al. 1991, 1997). Higher CO_2 levels would likely have resulted in higher plant productivity as well as higher water-use efficiency, thereby allowing a monsoon forest to mimic rainforest in terms of the spatio-temporal availability of fruit.

In conclusion, *Sivapithecus* appears to have inhabited a forest more seasonal than those great apes inhabit today, but a forest that was probably similar to modern ape habitat in terms of the spatio-temporal availability of fruit. This forest became fragmented over time as amount of annual rainfall decreased, resulting in the extinction of many forest frugivores, including *Sivapithecu*s.

Appendix 1 Microwear Data

Microwear data are listed for each tooth analyzed in this study. Fossil species include the hominoid *Sivapithecus* spp.; the suids *Propotamochoerus hysudricus* and *Hippopotamodon sivalense*; the tragulids *Dorcabune nagrii*, *Dorcatherium majus*, *Dorcatherium nagrii*, *Dorcatherium* "sp. D", and *Dorcatherium* "Y270 sp."; and the bovids "2-Keel", *Dorcadoxa porrecticornis*, *Elachistoceras khauristanensis*, a group of indeterminate small bovids, and *Tragoceridus* spp. Modern species collected specifically for this study include the hominoids *Pan troglodytes*, *Gorilla gorilla berengei*, *Gorilla gorilla gorilla*, and *Pongo pygmaeus*; and the suids *Babyrousa babyrussa* and *Hylochoerus meinertzhageni*. Characters include the following continuous variables: number of pits and number of scratches. Categorical variables include the presence/absence of the following: at least four large pits, at least four gouges, and at least four instances of cross scratches. Pits are defined as those features with a length:width ratio below 4:1, while scratches have ratios greater than 4:1. Gouges are very irregularly shaped large pits. Cross scratches are those running perpendicular to one another. In each of these categories, "Yes" for the presence of a character is indicated by a "1", while "no" is indicated by a "0". Finally, texture of scratches is characterized as predominantly fine, predominantly coarse, or a mixture of significant amounts of both fine and coarse. The texture assigned to each tooth is designated by a "1". At the bottom of the list of specimens analyzed for any given species, the average is calculated for number of pits and number of scratches. The percentage of specimens with large pits, gouges, and cross scratches is also calculated. Finally, the percentage for each category of scratch texture is calculated, with percentages summing to 100 percent across the three categories.

Microwear Data

Taxon	Spec.#	Age	Locality	# Pits	Lrg. Pits	Gouges	# Scr	Cross Sc	Fine	Coarse	Mixed
Sivapithecus	Y9986	10.0	Y311	25.5	1	0	17	1			1
	Y11533	10.0	Y311	25	1	1	15	1			1
	Y11998	10.0	Y311	32.5	1	1	30	1			1
	Y11999	10.0	Y311	26.5	1	1	13	1			1
	Y13566	9.3	Y350	29	1	1	21	1			1
	Y15000	9.3	Y410	33	1	0	21	1			1
	Y5001	9.3	Y207	20	1	1	20	1			1
	Y15030	9.3	Y416	29.5	1	1	17.5	1		1	
	Y5260	9.2	Y211	30	1	0	16	1			1
	Y9563	9.2	Y260	29.5	1	1	12.5	1			1
	Y9899	9.2	Y260	25	1	0	19.5	1			1
	Y9972	9.2	Y260	42.5	1	1	12	1			1
	Y9977	9.2	Y260	30.5	0	0	19	1			1
	Y13165	9.2	Y260	31	0	0	17	1		1	
	Y11704	9.2	Y317	18.5	0	1	20.5	1			1
	Y11707	9.2	Y317	16.5	1	1	27.5	1		1	
	Y11708	9.2	Y317	24.5	1	1	18	1			1
	Y4230	9.2	Y182	53.5	1	1	21.5	1			1
	Y4622	9.2	Y182	58.5	1	1	30	1			1
	Y16077		Y498	23	1	1	15	1			1
	AVG			**30.2**			**19.2**				
	STDEV			**10.5**			**5.2**				
	%				**85.0**	**70.0**		**100.0**	**0**	**15**	**85**
Propotamoch	Y17578	10.0	Y311	34	1	1	20.5	1			1
	Y22080	10.0	Y311	31.5	1	0	7	1	1		
	Y6226	9.5	Y228	26	1	1	17.5	1			1
	Y14044	9.5	Y262	23	1	0	27.5	1			1
	Y17088	9.5	Y262	49	1	1	13.5	1			1
	Y6011	9.3	Y341	35	1	1	11	1			1
	Y14501	9.3	Y269	23	1	1	19	1			1
	Y14502	9.3	Y269	28	1	0	19	1			1
	Y41368	9.3	Y269	23.5	1	1	14	1			1
	Y10224	9.3	Y309	37	1	1	15	1			1
	Y13587	9.3	Y227	21.5	1	1	8	1			1
	Y6137	9.3	Y221	19	1	1	18	1			1
	Y16132	9.3	Y310	22.5	1	0	19.5	1			1
	Y10998	9.2	Y314	27.5	1	1	20.5	1			1
	Y11007	9.2	Y314	26.5	1	1	19.5	1			1

Microwear Data

Taxon	Spec.#	Age	Locality	# Pits	Lrg. Pits	Gouges	# Scr	Cross Sc	Fine	Coarse	Mixed
Propotamoch	Y5112	9.2	Y211	27	1	0	11.5	1	1		
	Y9355	9.2	Y260	36	1	1	20	1			1
	Y4554	9.2	Y182	55	1	0	19	1			1
	Y7654	9.2	Y182	47	1	1	16	1			1
	Y13678	8.9	Y367	58.5	1	1	22.5	1			1
	Y4599	8.9	Y193	41.5	1	1	19	1			1
	Y17968	8.5	Y604	20.5	1	1	17	1			1
	Y4813	8.5	Y203	12	1	0	19	1		1	
	Y457	8.1	Y024	24.5	1	1	20.5	1		1	
	Y49515	8.0	Y599	26	1	0	19.5	1		1	
	Y50880	7.8	Y949	34.5	1	0	21.5	1			1
	Y49840	7.8	Y906	39	1	1	8.5	1			1
	Y49844	7.8	Y906	28	1	1	23	1			1
	Y18124	7.2	Y452	22	1	0	15.5	1			1
	Y14045	7.1	Y370	11.5	1	1	13	1			1
	Y49816	7.0	Y916	32.5	1	0	14.5	1	1		
	Y85		YO10	42.5	1	1	23	1			1
	Y2807		Y105	42	0	0	18.5	1			1
	Y19698			31.5	1	0	18.5	1			1
	Y19748		L37	30.5	1	0	24	1			1
	Y19803			24.5	1	1	20.5	1			1
	Y51575		Y539	27.5	1	1	19	1			1
	Y51848		Y269	27	1	0	20	1			1
	Y52623		Y960	37.5	1	1	20	1			1
	Y52855		O84	30.5	1	0	7.5	1	1		
	D3		D005	24.5	1	1	11	1			1
	AVG			**30.7**			**17.4**				
	SD			**10.2**			**4.7**				
	%				**97.6**	**61.0**		**100.0**	**9.8**	**7.3**	**82.9**
Hippopot	Y18261	10.4	Y454	29	1	0	25.5	1			1
	Y18264	10.4	Y454	37	1	0	20.5	1			1
	Y18276	10.4	Y454	28	0	0	10.5	1			1
	Y18280	10.4	Y454	42	1	1	16	1			1
	D36	10.4	D27	20	1	1	11.5	1			1
	Y53009	10.2	Y450	13.5	0	0	21	1			1
	Y41401	10.0	Y311	36	1	0	18	1			1
	Y13972	10.0	Y251	56	1	1	14	1		1	
	Y16049	10.0	Y251	28.5	1	0	23	1			1

Microwear Data

Taxon	Spec.#	Age	Locality	# Pits	Lrg. Pits	Gouges	# Scr	Cross Sc	Fine	Coarse	Mixed
Hippopot	Y46568	10.0	Y251	23.5	1	1	16	1		1	
	Y12861	9.3	Y309	28	1	1	21.5	1			1
	Y4984	9.3	Y159	30.5	1	0	28	1			1
	Y5276	9.2	Y211	11	1	0	20.5	1			1
	Y13524	9.2	Y239	25.5	0	0	18	1		1	
	Y14112	9.2	Y260	31	1	1	15	1			1
	Y27915	9.2	Y260	43	1	1	23.5	1			1
	Y16085	9.2	Y182	28.5	1	1	21	1	1		
	Y17937	8.5	KL12	12.5	0	0	17	1			1
	Y4086	8.3	Y174	26	1	1	16	1			1
	Y330	8.0	YO17	29.5	0	0	16.5	1			1
	Y50019	7.8	Y917	27	1	0	19.5	1			1
	Y50457	7.8	Y947	9	0	0	17.5	1			1
	Y50134	7.2	Y932	30	1	0	22	1			1
	AVG			**28.0**			**18.8**				
	STDEV			**10.8**			**4.3**				
	%				**73.9**	**39.1**		**100.0**	**4.3**	**13.0**	**82.6**
Dorcab. nagrii	Y726	8.1	Y036	37.5	1	1	14	1			1
	Y6857	9.3	Y227	30	1	0	21	1		1	
	Y10220	9.3	Y309	35	1	0	13	0		1	
	Y11499	10.0	Y311	27.5	1	1	13.5	1			1
	Y14351	10.2	Y450	29.5	1	0	7.5	0			1
	Y14867	10.3	Y395	28.5	1	1	8	1		1	
	Y15568	9.6	Y462	33.5	1	1	17.5	1			1
	Y16071	10.4	Y259	16	1	1	23.5	1			1
	Y30743	10.0	Y311	33.5	1	0	28	1			1
	Y51962		Y1005	33	1	0	12	1			1
	AVG			**30.4**			**15.8**				
	STDEV			**5.9**			**6.7**				
	%				**100.0**	**50.0**		**80.0**	**0.0**	**30.0**	**70.0**
D. majus	D482	10.4	D027	39	1	1	17	1			1
	Y15126	10.1	Y450	22.5	1	1	14.5	0			1
	Y16847	10.1	Y450	21	1	1	17	1			1
	Y18007	10.1	Y450	17.5	1	0	18.5	1			1
	Y50639	10.1	Y450	24	1	1	21	0			1
	Y50647	10.1	Y450	31	1	1	18.5	0		1	

Microwear Data

Taxon	Spec.#	Age	Locality	# Pits	Lrg. Pits	Gouges	# Scr	Cross Sc	Fine	Coarse	Mixed
D. majus	Y11368	10.0	Y311	19.5	1	0	18.5	1			1
	Y11391	10.0	Y311	30.5	1	0	31	1			1
	Y17707	10.0	Y311	19.5	0	0	25	1			1
	Y17748	10.0	Y311	21	0	0	15	1			1
	Y19949	10.0	Y311	28.5	1	0	13	0	1		
	Y20169	10.0	Y311	12	0	1	20	1			1
	Y20552	10.0	Y311	27	1	1	17.5	1			1
	Y22082	10.0	Y311	23	1	0	10	0			1
	Y22083	10.0	Y311	17	1	0	22.5	1	1		
	Y27827	10.0	Y311	30.5	1	0	15	1			1
	Y41417	10.0	Y311	33	1	0	15	1	1		
	Y41418	10.0	Y311	33	1	1	11.5	0			1
	Y47503	10.0	Y311	30.5	1	0	20.5	1			1
	Y47533	10.0	Y311	25.5	1	0	23.5	1			1
	Y47604	10.0	Y311	11.5	1	0	22	1			1
	Y47671	10.0	Y311	27.5	1	0	14.5	1			1
	Y47680	10.0	Y311	22.5	0	0	11.5	1			1
	Y5567	9.3	Y310	46	1	0	20.5	1			1
	Y49530	8.0	Y888	27	1	1	13	1			1
	Y49555	8.0	Y539	44.5	1	0	16	0			1
	Y16236	8.0	Y540	30	1	0	22	1			1
	Y50441	7.8	Y947	30	1	0	12	0			1
	Y52556	7.3	Y941	38.5	1	0	16	0			1
	Y19707			56	1	0	11	1			1
	Y52833		Y981	13.5	0	0	22.5	1	1		
	Y52856		L084	37	0	0	9	0	1		
	AVG			**27.8**			**17.3**				
	STDEV			**10.0**			**4.9**				
	%				**81.3**	**28.1**		**68.8**	**15.6**	**3.1**	**81.3**
Dorcath. nagrii	Y17619	9.2	Y260	22	1	0	9	0			1
	Y13771	8.7	Y166	29	1	0	12	0	1		
	Y13772	8.7	Y166	15	0	0	9	0	1		
	Y17950	8.5	Y604	32	1	0	23	1			1
	Y16204	8.4	Y535	30	1	0	9	1	1		
	Y16205	8.4	Y535	23	1	0	13	1	1		
	Y16207	8.4	Y535	30	1	0	8	0			1
	Y18396	8.1	Y605	25	1	1	8	0			1

Microwear Data

Taxon	Spec.#	Age	Locality	# Pits	Lrg. Pits	Gouges	# Scr	Cross Sc	Fine	Coarse	Mixed
Dorcath. nagrii	Y49544	7.9	Y399	31	0	0	8	0	1		
	Y16402	7.9	KL11	26	1	1	10	0			1
	Y51936		Y992	32	1	0	15	0		1	
	AVG			**26.8**			**11.3**				
	STDEV			**5.3**			**4.5**				
	%				**81.8**	**18.2**		**27.3**	**45.5**	**9.1**	**45.5**
D. "sp. D"	Y14350	10.2	Y450	31.5	1	0	13	0			1
	Y16837	10.2	Y450	38	1	0	12	0			1
	Y18005	10.2	Y450	36.5	1	0	15.5	1			1
	Y18006	10.2	Y450	39.5	1	1	20	1			1
	Y6358	10.0	Y311	29.5	1	0	17	0			1
	Y12050	10.0	Y311	45.5	1	1	25.5	1			1
	Y17580	10.0	Y311	36.5	1	0	11.5	1			1
	Y17705	10.0	Y311	21.5	1	1	23.5	1			1
	Y20127	10.0	Y311	46	1	1	13.5	1			1
	Y22087	10.0	Y311	47.5	1	0	12	0		1	
	Y22158	10.0	Y311	34	0	0	18	0			1
	Y26227	10.0	Y311	46	1	1	12	0			1
	Y41425	10.0	Y311	36.5	1	0	19	1			1
	Y46284	10.0	Y311	36	1	0	17	1	1		
	Y47591	10.0	Y311	22	0	0	12	0			1
	Y47658	10.0	Y311	27.5	1	0	17.5	1			1
	Y47686	10.0	Y311	27	1	0	10	0	1		
	Y13671	9.3	Y350	34	1	1	13	0	1		
	Y10777	9.3	Y310	27	1	0	14.5	0			1
	Y10780	9.3	Y310	61	1	1	19.5	1			1
	Y11110	9.2	Y314	56	1	1	18.5	1			1
	Y4603	8.9	Y193	41.5	1	0	22.5	1			1
	Y47116	8.4	Y535	40.5	1	0	15.5	1			1
	Y47117	8.4	Y535	15	1	0	17	1			1
	Y5049	8.3	Y028	19	1	0	15	1			1
	Y236	8.0	Y017	34.5	1	0	12	0	1		
	Y50210	7.3	Y935	23	1	1	10.5	1			1
	Y50997	7.3	Y935	21.5	1	0	9.5	0			1
	Y50999	7.3	Y935	26.5	1	0	15.5	1		1	
	Y52868	7.3	Y935	20.5	1	0	14.5	1			1
	Y50403	7.3	Y941	33.5	1	0	9.5	1	1		

Microwear Data

Taxon	Spec.#	Age	Locality	# Pits	Lrg. Pits	Gouges	# Scr	Cross Sc	Fine	Coarse	Mixed
D. "sp. D"	**AVG**			**34.0**			**15.4**				
	STDEV			**10.9**			**4.1**				
	%				**93.5**	**29.0**		**61.3**	**16.1**	**6.5**	**77.4**
Dorcath. "Y270"	Y7458	9.3	Y245	18	1	0	10	0			1
	Y15576	9.6	Y426	33	1	0	12	0			1
	AVG			**25.5**			**11.0**				
	STDEV			**10.6**			**1.4**				
	%				**100.0**	**0.0**		**0.0**	**0.0**	**0.0**	**100.0**
"2-Keel"	Y50356	7.3	Y941	14.5	1	0	8.5	1		1	
	Y50399	7.3	Y941	15	0	0	5	0			1
	AVG			**14.8**			**6.8**				
	STDEV			**0.4**			**2.5**				
	%				**50.0**	**0.0**		**50.0**	**0.0**	**50.0**	**50.0**
Dorcadoxa	Y2817	8.0	Y017	19	1	0	25.5	1			1
	Y12367	8.9	Y325	20	0	0	17	0			1
	Y50737	6.4	D013	22	1	0	20	1			1
	Y50771	7.3	Y927	23	0	0	14	1	1		
	Y50824	7.2	Y452	29	0	0	6.5	0	1		
	AVG			**22.6**			**16.6**				
	STDEV			**3.9**			**7.1**				
	%				**40.0**	**0.0**		**60.0**	**40.0**	**0.0**	**60.0**
Elachistoceras	Y17090	9.5	Y262	19.5	0	0	9.5	0	1		
	Y18346	8.5	KL12	26	0	0	5	0	1		
	AVG			**22.8**			**7.3**				
	STDEV			**4.6**			**3.2**				
	%				**0.0**	**0.0**		**0.0**	**100.0**	**0.0**	**0.0**
Indet. Sm. bov.	Y16836	10.2	Y450	15	0	0	11	0	1		
	Y46345	9.9	Y870	19	1	0	10.5	0	1		
	Y46348	9.9	Y870	31.5	1	0	4.5	0	1		
	Y14860	8.7	Y388	32	1	0	8	0	1		

Microwear Data

Taxon	Spec.#	Age	Locality	# Pits	Lrg. Pits	Gouges	# Scr	Cross Sc	Fine	Coarse	Mixed
Indet. Sm. bov.	Y542	8.1	Y024	10	0	0	16	0	1		
	Y14860	8.7	Y388	32	1	0	8	0	1		
	Y16978	8.0	Y545	34	1	0	17	1	1		
	Y14089	7.1	Y374	32	0	0	20.5	0			1
	AVG			**24.8**			**12.5**				
	STDEV			**9.8**			**5.6**				
	%				**57.1**	**0.0**		**14.3**	**85.7**	**0.0**	**14.3**
Tragoceridus	Y11593	9.3	Y312	26	0	0	13	0			1
	Y5127	9.2	Y211	47	1	0	11	1		1	
	Y15012	8.9	Y476	27	0	0	7.5	1	1		
	Y14882	8.6	Y389	21	1	0	16.5	1			1
	Y2349	8.1	Y024	24	0	0	8.5	0			1
	Y18149	6.4	D013	19.5	1	1	26.5	1			1
	Y18210	6.3	Y453	20	1	0	19	0		1	
	Y9196		Y256	26	0	0	7	0			1
	AVG			**26.3**			**13.6**				
	STDEV			**8.9**			**6.7**				
	%				**50.0**	**12.5**		**50.0**	**12.5**	**25.0**	**62.5**
Pan troglodytes	46416		MCZ	22	1	0	14.5	1			1
	19187		MCZ	19.5	1	0	15	1			1
	17702		MCZ	31	1	1	14	1			1
	20041		MCZ	19	1	1	15	1			1
	9493		MCZ	33	1	1	15	1			1
	9494		MCZ	24.5	1	1	6.5	1			1
	23163		MCZ	30	1	1	14.5	1	1		
	23164		MCZ	37	1	1	7.5	1	1		
	23167		MCZ	24	1	1	18.5	1			1
	38018		MCZ	14	0	0	10.5	1	1		
	38019		MCZ	19	1	1	16.5	1			1
	38020		MCZ	30	1	1	11.5	1			1
	6244		MCZ	46	1	1	12.5	1	1		
	15312		MCZ	41.5	1	1	19	1			1
	AVG			**27.9**			**13.6**				
	STDEV			**9.3**			**3.6**				
	%				**92.9**	**78.6**		**100.0**	**28.6**	**0.0**	**71.4**

Microwear Data

Taxon	Spec.#	Age	Locality	# Pits	Lrg. Pits	Gouges	# Scr	Cross Sc	Fine	Coarse	Mixed
G. g. berengei	38017		MCZ	14	0	0	29	1			1
	23182		MCZ	26.5	1	1	38	1			1
	AVG			**20.3**			**33.5**				
	%				**50.0**	**50.0**		**100.0**	**0.0**	**0.0**	**100.0**
G. g. gorilla	23162		MCZ	16	0	0	21.5	1			1
	20038		MCZ	27.5	1	1	11	0			1
	29049		MCZ	18.5	1	0	20	1			1
	38326		MCZ	33.5	1	1	16	1			1
	29048		MCZ	17.5	1	0	35.5	1			1
	20043		MCZ	29	1	1	24.5	1			1
	37262		MCZ	18.5	1	0	10	1			1
	37266		MCZ	17.5	1	1	17.5	1			1
	37261		MCZ	9	0	0	18	1			1
	62393		MCZ	17.5	0	0	25	1	1		
	20039		MCZ	39	1	1	19	1	1		
	23160		MCZ	26.5	1	0	17.5	1			1
	23161		MCZ	17.5	0	0	17.5	1		1	
	26850		MCZ	19.5	1	1	19.5	1		1	
	AVG			**21.9**			**19.5**				
	STDEV			**8.0**			**6.2**				
	%				**71.4**	**42.9**		**92.9**	**14.3**	**14.3**	**71.4**
Pongo pygmaeus	37363		MCZ	25	1	0	17.5	1			1
	37518		MCZ	22	1	0	18	1			1
	37519		MCZ	43	1	1	19	1			1
	37516		MCZ	22.5	1	1	29.5	1			1
	50960		MCZ	33.5	1	1	21	1			1
	5061		MCZ	39.5	1	0	17	1			1
	37362		MCZ	14	0	0	26	1			1
	37517		MCZ	41.5	1	1	13.5	1			1
	16140		Borneo	35.5	1	1	17.5	1		1	
	AVG			**30.7**			**19.9**				
	STDEV			**10.2**			**4.9**				
	%				**88.9**	**55.6**		**100.0**	**0.0**	**11.1**	**88.9**

Microwear Data

Taxon	Spec.#	Age	Locality	# Pits	Lrg. Pits	Gouges	# Scr	Cross Sc	Fine	Coarse	Mixed
Babyrousa	1913		MCZ	19	1	1	26.5	1		1	
	1910		MCZ	28.5	1	1	31	1			1
	46401		MCZ	18	1	0	22	1			1
	1912		MCZ	20	1	0	27.5	1			1
	46400		MCZ	34.5	1	0	19.5	1			1
	9399		MCZ	37.5	1	1	26.5	1		1	
	47389		MCZ	30	1	1	27	1		1	
	9401		MCZ	24	1	0	27	1			1
	AVG			**26.4**			**25.9**				
	ST DEV			**7.4**			**3.5**				
	%				**100.0**	**50.0**		**100.0**	**0.0**	**37.5**	**62.5**
Hylochoerus	27851		MCZ	22	1	0	16.5	1			1
	21202		MCZ	22	1	0	17.5	1	1		
	39429		MCZ	26.5	0	0	17.5	1	1		
	38011		MCZ	16.5	0	0	15.5	1	1		
	AVG			**21.8**			**16.8**				
	ST DEV			**4.1**			**1.0**				
	%				**50.0**	**0.0**		**100.0**	**75.0**	**0.0**	**25.0**

Appendix 2 Equid Intra-tooth Profiles

Carbon and oxygen stable isotope ratios (PDB) in 15 *Hipparion* teeth sampled throughout the length of the tooth. Samples begin at the crown of the tooth, are numbered consecutively, and end at the cervical margin. Missing values represent unsuccessful analyses on the mass spectrometer (five out of 243 samples). For tooth position, "M" represents "molar" and "P" represents "premolar". Three-point averages were taken to smooth carbon and oxygen curves, and the resulting values are listed as well.

Equid Intra-tooth Profiles

Specimen	Age (Ma)	Locality	Tooth	$\delta^{13}C$	$\delta^{18}O$	3-pt. Avg. $\delta^{13}C$	3-pt. Avg. $\delta^{18}O$
1-1Y9488	10.1	Y258	M3	-8.4	0.3	#N/A	#N/A
1-2 Y9488				-8.0	0.3	#N/A	#N/A
1-3 Y9488				-7.8	-0.7	-8.0	0.0
1-4 Y9488				-7.5	-1.1	-7.8	-0.5
1-5 Y9488				-7.9	-1.6	-7.7	-1.1
1-6 Y9488				-8.8	-2.2	-8.1	-1.6
1-7 Y9488				-9.4	-2.0	-8.7	-1.9
1-8 Y9488				-9.8	-1.1	-9.3	-1.7
1-9 Y9488				-9.7	0.3	-9.6	-0.9
1-10 Y9488				-9.7	-0.5	-9.7	-0.4
1-11 Y9488						-9.7	-0.1
1-12 Y9488				-9.6	-0.4	-9.7	-0.5
1-13 Y9488				-9.2	-0.9	-9.4	-0.6
1-14 Y9488				-8.8	-0.9	-9.2	-0.7
1-15 Y9488				-8.6	0.0	-8.9	-0.6
1-16 Y9488				-8.2	-1.6	-8.5	-0.8
16-1 Y11881	10	Y311	M3	-11.4	-6.1	#N/A	#N/A
16-2 Y11881				-11.7	-6.8	#N/A	#N/A
16-3 Y11881				-11.7	-6.5	-11.6	-6.5
16-4 Y11881				-11.9	-6.2	-11.8	-6.5
16-5 Y11881				-11.9	-5.5	-11.8	-6.1
16-6 Y11881				-11.7	-5.6	-11.8	-5.8
16-7 Y11881				-11.9	-5.0	-11.8	-5.4
16-8 Y11881				-11.8	-5.0	-11.8	-5.2
16-9 Y11881				-12.1	-4.6	-12.0	-4.9
16-10 Y11881				-11.9	-4.1	-12.0	-4.6
16-11 Y11881				-11.6	-3.9	-11.9	-4.2
16-12 Y11881				-11.7	-3.9	-11.8	-4.0
16-13 Y11881				-11.6	-4.1	-11.7	-4.0
16-14 Y11881				-11.5	-4.4	-11.6	-4.1
16-15 Y11881				-11.8	-5.3	-11.6	-4.6
16-16 Y11881				-11.3	-6.2	-11.5	-5.3
16-17 Y11881				-11.4	-6.3	-11.5	-5.9
16-18 Y11881				-11.5	-6.1	-11.4	-6.2
7-1 Y46669	9.7	Y875	M3	-12.4	-4.3	#N/A	#N/A
7-2 Y46669				-12.5	-3.9	#N/A	#N/A
7-3 Y46669				-12.7	-4.0	-12.6	-4.1
7-4 Y46669				-12.6	-4.3	-12.6	-4.1

Equid Intra-tooth Profiles

Specimen	Age (Ma)	Locality	Tooth	$\delta^{13}C$	$\delta^{18}O$	3-pt. Avg. $\delta^{13}C$	3-pt. Avg. $\delta^{18}O$
7-5 Y46669				-12.7	-4.8	-12.7	-4.4
7-6 Y46669				-12.6	-4.7	-12.6	-4.6
7-7 Y46669				-12.5	-4.3	-12.6	-4.6
7-8 Y46669				-12.5	-5.3	-12.5	-4.8
7-9 Y46669				-12.5	-4.9	-12.5	-4.8
7-10 Y46669				-12.7	-5.3	-12.5	-5.2
7-11 Y46669						-12.6	-5.1
7-12 Y46669				-12.5	-6.1	-12.6	-5.7
7-13 Y46669				-12.4	-5.8	-12.5	-6.0
7-14 Y46669				-12.1	-6.0	-12.4	-6.0
7-15 Y46669				-12.2	-5.2	-12.3	-5.7
7-16 Y46669				-12.5	-4.7	-12.3	-5.3
2-1 Y19810	9.3	Y312	M3			#N/A	#N/A
2-2 Y19810				-9.6	-1.0	#N/A	#N/A
2-3 Y19810				-7.6	-0.4	-8.6	-0.7
2-4 Y19810				-9.4	-0.5	-8.9	-0.6
2-5 Y19810				-9.3	-1.7	-8.8	-0.9
2-6 Y19810				-8.9	-1.1	-9.2	-1.1
2-7 Y19810				-9.0	-1.0	-9.1	-1.3
2-8 Y19810				-9.0	-1.1	-9.0	-1.1
2-9 Y19810				-9.4	-1.7	-9.1	-1.3
2-10 Y19810				-9.3	-1.6	-9.3	-1.5
2-11 Y19810				-8.8	-2.0	-9.2	-1.7
2-13 Y19810				-8.5	-2.5	-8.9	-2.0
2-14 Y19810				-9.4	-2.4	-8.9	-2.3
2-15 Y1810				-9.9	-2.8	-9.2	-2.6
2-16 Y19810				-9.9	-2.1	-9.7	-2.5
2-17 Y19810				-9.9	-2.0	-9.9	-2.3
8-1 Y5097	9	Y210	M1			#N/A	#N/A
8-2 Y5097				-11.1	-1.4	#N/A	#N/A
8-3 Y5097				-10.9	-1.7	-11.0	-1.5
8-4 Y5097				-10.9	-2.1	-11.0	-1.7
8-5 Y5097				-11.4	-2.3	-11.1	-2.0
8-6 Y5097				-10.9	-2.4	-11.1	-2.3
8-7 Y5097				-10.9	-3.6	-11.1	-2.8
8-8 Y5097				-10.9	-3.3	-10.9	-3.1
8-9 Y5097				-10.6	-3.3	-10.8	-3.4
8-10 Y5097				-10.9	-4.2	-10.8	-3.6

Equid Intra-tooth Profiles

Specimen	Age (Ma)	Locality	Tooth	$\delta^{13}C$	$\delta^{18}O$	3-pt. Avg. $\delta^{13}C$	3-pt. Avg. $\delta^{18}O$
8-11 Y5097				-11.2	-4.0	-10.9	-3.8
8-12 Y5097				-11.3	-4.5	-11.2	-4.2
8-13 Y5097				-10.5	-3.2	-11.0	-3.9
8-14 Y5097				-11.6	-4.8	-11.1	-4.2
8-15 Y5097				-11.6	-4.3	-11.2	-4.1
4-1 Y3699	8.7	Y158	P3	-3.2	-3.4	#N/A	#N/A
4-2 Y3699				-3.5	-3.1	#N/A	#N/A
4-3 Y3699				-3.7	-2.9	-3.5	-3.1
4-4 Y3699				-4.2	-2.9	-3.8	-3.0
4-5 Y3699				-4.4	-2.5	-4.1	-2.8
4-6 Y3699				-4.3	-1.6	-4.3	-2.3
4-7 Y3699				-4.4	-1.3	-4.4	-1.8
4-8 Y3699				-3.9	0.0	-4.2	-1.0
4-9 Y3699				-4.2	-0.1	-4.1	-0.5
4-10 Y3699				-3.8	0.0	-4.0	-0.1
4-11 Y3699				-3.4	-0.5	-3.8	-0.2
4-12 Y3699				-3.4	-0.2	-3.5	-0.2
4-13 Y3699				-3.1	-0.5	-3.3	-0.4
11-1 Y15610	8.5	Y467	M1	-9.2	-4.7	#N/A	#N/A
11-2 Y15610				-9.3	-5.2	#N/A	#N/A
11-3 Y15610				-9.3	-5.3	-9.3	-5.1
11-4 Y15610				-9.5	-5.9	-9.4	-5.5
11-5 Y15610				-9.6	-6.1	-9.5	-5.8
11-6 Y15610				-10.0	-5.9	-9.7	-5.9
11-7 Y15610				-10.2	-5.9	-9.9	-5.9
11-8 Y15610				-10.2	-5.7	-10.1	-5.8
11-9 Y15610				-10.4	-5.2	-10.3	-5.6
11-10 Y15610				-10.6	-4.8	-10.4	-5.2
11-11 Y15610				-10.5	-4.7	-10.5	-4.9
11-12 Y15610				-10.4	-4.4	-10.5	-4.6
11-13 Y15610				-10.4	-3.8	-10.4	-4.3
11-14 Y15610				-10.3	-3.9	-10.4	-4.0
11-15 Y15610				-10.2	-4.5	-10.3	-4.1
14-1 Y15403	8.4	Y443	P3	-7.5	-2.3	#N/A	#N/A
14-2 Y15403				-7.5	-2.2	#N/A	#N/A
14-3 Y15403				-7.7	-2.3	-7.6	-2.3
14-4 Y15403				-7.3	-2.7	-7.5	-2.4

Equid Intra-tooth Profiles

Specimen	Age (Ma)	Locality	Tooth	$\delta^{13}C$	$\delta^{18}O$	3-pt. Avg. $\delta^{13}C$	3-pt. Avg. $\delta^{18}O$
14-5 Y15403				-7.3	-3.0	-7.5	-2.7
14-6 Y15403				-7.8	-3.1	-7.5	-2.9
14-7 Y15403				-8.3	-3.4	-7.8	-3.2
14-8 Y15403				-8.7	-2.9	-8.3	-3.1
14-9 Y15403				-9.0	-2.8	-8.6	-3.0
14-10 Y15403				-8.8	-2.0	-8.8	-2.5
14-11 Y15403				-8.7	-1.4	-8.8	-2.1
14-12 Y15403				-8.2	-0.9	-8.5	-1.4
14-13 Y15403				-7.5	-1.2	-8.1	-1.2
14-14 Y15403				-6.9	-0.7	-7.5	-0.9
3-1 Y3050	8.3	Y113	P4	-5.0	-3.6	#N/A	#N/A
3-2 Y3050				-4.6	-3.9	#N/A	#N/A
3-3 Y3050				-4.5	-4.0	-4.7	-3.9
3-4 Y3050				-4.6	-3.9	-4.6	-3.9
3-5 Y3050				-4.9	-4.2	-4.7	-4.0
3-6 Y3050				-4.6	-4.4	-4.7	-4.1
3-7 Y3050				-4.7	-4.7	-4.7	-4.4
3-8 Y3050				-5.1	-5.3	-4.8	-4.8
3-9 Y3050				-5.6	-5.0	-5.2	-5.0
3-10 Y3050				-5.7	-5.7	-5.5	-5.3
3-11 Y3050				-5.9	-5.9	-5.7	-5.5
3-12 Y3050				-6.7	-5.5	-6.1	-5.7
3-13 Y3050				-7.2	-5.3	-6.6	-5.6
3-14 Y3050				-7.8	-4.5	-7.2	-5.1
3-15 Y3050				-7.8	-3.8	-7.6	-4.5
3-16 Y3050				-8.1	-3.8	-7.9	-4.0
3-17 Y3050				-8.1	-3.8	-8.0	-3.8
3-18 Y3050				-7.7	-4.1	-8.0	-3.9
3-19 Y3050				-7.6	-4.0	-7.8	-4.0
13-1 Y49548	8	Y890	M2	-6.1	-0.7	#N/A	#N/A
13-2 Y49548				-6.2	-0.7	#N/A	#N/A
13-3 Y49548				-6.3	-1.2	-6.2	-0.8
13-4 Y49548				-6.3	-0.9	-6.3	-0.9
13-5 Y49548				-6.8	-1.0	-6.5	-1.0
13-6 Y49548				-7.3	-0.9	-6.8	-0.9
13-7 Y49548				-7.6	-0.1	-7.2	-0.7
13-8 Y49548				-7.8	-0.5	-7.6	-0.5
13-9 Y49548				-7.6	-0.4	-7.7	-0.3

Equid Intra-tooth Profiles

Specimen	Age (Ma)	Locality	Tooth	$\delta^{13}C$	$\delta^{18}O$	3-pt. Avg. $\delta^{13}C$	3-pt. Avg. $\delta^{18}O$
13-10 Y49548				-7.5	-0.5	-7.6	-0.5
13-11 Y49548				-7.2	-0.4	-7.4	-0.4
13-12 Y49548				-7.3	-0.4	-7.3	-0.4
13-13 Y49548				-7.0	0.0	-7.2	-0.3
13-14 Y49548				-6.6	0.5	-7.0	0.0
13-15 Y49548				-6.3	0.6	-6.6	0.4
13-16 Y49548				-6.2	-0.2	-6.4	0.3
13-17 Y49548				-6.0	-1.6	-6.2	-0.4
13-18 Y49548				-5.6	-1.8	-5.9	-1.2
13-19 Y49548				-5.2	-1.0	-5.6	-1.5
13-20 Y49548				-5.0	-1.0	-5.3	-1.3
13-21 Y49548						-5.1	-1.0
13-22 Y49548				-5.7	-2.7	-5.4	-1.9
13-23 Y49548				-6.4	-3.2	-6.1	-3.0
5-1 Y579	7.9	Y029	M2	-1.3	-1.5	#N/A	#N/A
5-2 Y579				-1.3	-1.3	#N/A	#N/A
5-3 Y579				-1.1	-0.6	-1.2	-1.1
5-4 Y579				-1.4	-0.4	-1.3	-0.8
5-5 Y579				-1.6	-0.1	-1.4	-0.4
5-6 Y579				-1.5	-0.5	-1.5	-0.4
5-7 Y579				-1.3	-0.6	-1.5	-0.4
5-8 Y579				-1.1	-0.2	-1.3	-0.4
5-9 Y579				-1.2	-0.4	-1.2	-0.4
5-10 Y579				-1.5	-1.0	-1.3	-0.5
5-11 Y579				-1.3	0.5	-1.3	-0.3
5-12 Y579				-1.2	0.0	-1.3	-0.2
5-13 Y579				-1.5	-0.6	-1.4	0.0
5-14 Y579				-1.6	-1.1	-1.5	-0.5
5-15 Y579				-1.8	-0.5	-1.7	-0.7
5-16 Y579				-2.0	-0.6	-1.8	-0.7
12-1 Y15293	7.7	Y435	P3	-3.7	0.4	#N/A	#N/A
12-2 Y15293				-4.3	-1.0	#N/A	#N/A
12-3 Y15293				-4.1	-0.6	-4.0	-0.4
12-4 Y15293				-4.6	-0.3	-4.3	-0.6
12-5 Y15293				-5.0	-0.5	-4.6	-0.5
12-6 Y15293				-5.4	0.7	-5.0	0.0
12-7 Y15293				-5.9	0.1	-5.5	0.1

Equid Intra-tooth Profiles

Specimen	Age (Ma)	Locality	Tooth	$\delta^{13}C$	$\delta^{18}O$	3-pt. Avg. $\delta^{13}C$	3-pt. Avg. $\delta^{18}O$
12-8 Y15293				-6.1	1.0	-5.8	0.6
12-9 Y15293				-6.1	0.9	-6.0	0.7
12-10 Y15293				-5.9	0.6	-6.0	0.8
12-11 Y15293				-4.5	2.3	-5.5	1.3
12-12 Y15293				-5.0	1.4	-5.2	1.4
12-13 Y15293				-4.9	0.1	-4.8	1.3
12-14 Y15293				-4.2	0.3	-4.7	0.6
6-1 Y50130	7.2	Y932	M3	-6.6	1.6	#N/A	#N/A
6-2 Y50130				-6.3	1.5	#N/A	#N/A
6-3 Y50130				-6.0	1.1	-6.3	1.4
6-4 Y50130				-5.8	1.3	-6.0	1.3
6-5 Y50130				-5.6	1.3	-5.8	1.2
6-6 Y50130				-5.6	1.8	-5.7	1.5
6-7 Y50130				-5.8	1.9	-5.7	1.7
6-8 Y50130				-5.5	1.8	-5.6	1.8
6-9 Y50130				-5.8	1.7	-5.7	1.8
6-10 Y50130				-5.6	1.4	-5.6	1.6
6-11 Y50130				-5.5	1.0	-5.6	1.3
6-12 Y50130				-5.5	1.2	-5.5	1.2
6-13 Y50130				-5.0	-0.2	-5.3	0.7
6-14 Y50130				-5.0	0.8	-5.2	0.6
6-15 Y50130				-5.9	0.2	-5.3	0.2
9-1 Y50803	6.9	Y925	P4	-6.2	-2.7	#N/A	#N/A
9-2 Y50803				-6.1	-2.0	#N/A	#N/A
9-3 Y50803				-6.0	-2.1	-6.1	-2.3
9-4 Y50803				-5.7	-1.1	-5.9	-1.7
9-5 Y50803				-5.4	-1.0	-5.7	-1.4
9-6 Y50803				-5.9	-1.4	-5.7	-1.2
9-7 Y50803				-6.8	-2.0	-6.1	-1.5
9-8 Y50803				-7.1	-1.8	-6.6	-1.7
9-9 Y50803				-7.5	-2.0	-7.1	-1.9
9-10 Y50803				-7.6	-2.1	-7.4	-2.0
9-11 Y50803				-7.4	-1.9	-7.5	-2.0
9-12 Y50803				-7.2	-1.5	-7.4	-1.8
9-13 Y50803				-7.1	-1.1	-7.2	-1.5
9-14 Y50803				-6.8	-1.7	-7.0	-1.5
9-15 Y50803				-6.9	-1.8	-6.9	-1.5

Equid Intra-tooth Profiles

Specimen	Age (Ma)	Locality	Tooth	$\delta^{13}C$	$\delta^{18}O$	3-pt. Avg. $\delta^{13}C$	3-pt. Avg. $\delta^{18}O$
9-16 Y50803				-7.5	-0.9	-7.1	-1.5
9-17 Y50803				-7.0	-0.4	-7.1	-1.0
10-1 Y49784	6.3	Y912	M1	-1.9	2.0	#N/A	#N/A
10-2- Y49784				-2.0	0.7	#N/A	#N/A
10-3 Y49784				-2.3	1.3	-2.1	1.3
10-4 Y49784				-2.6	1.8	-2.3	1.3
10-5 Y49784				-3.0	2.0	-2.6	1.7
10-6 Y49784				-3.3	1.7	-3.0	1.8
10-7 Y49874				-3.6	2.0	-3.3	1.9
10-8 Y49874				-2.6	1.3	-3.1	1.7
10-9 Y49874				-4.5	1.1	-3.6	1.5
10-10 Y49874				-4.8	1.1	-4.0	1.2
10-11Y49874				-4.6	0.6	-4.6	0.9
10-12 Y49874				-4.2	0.1	-4.5	0.6
10-13 Y49874				-4.2	0.4	-4.3	0.4
10-14 Y49874				-4.5	1.7	-4.3	0.7
10-15 Y49874				-3.7	-0.9	-4.1	0.4

References

Anderson, D., and D. Archer
2002 Glacial-interglacial stability of ocean pH inferred from foraminifer dissolution rates. *Nature* 416:70–73.

Araguas-Araguas, L., and K. Froehlich
1998 Stable isotope composition or precipitation over southeast Asia. *Journal of Geophysical Research* 103:28721–28742.

Ayliffe, L., and A. Chivas
1990 Oxygen isotope composition of the bone phosphate of Australian kangaroos: potential as a paleoenvironmental recorder. *Geochimica et Cosmochimica Acta* 54:2603–2609.

Ayliffe, L. K., A. R. Chivas, and M. G. Leakey
1994 The retention of primary oxygen isotope composition of fossil elephant skeletal phosphate. *Geochimica et Cosmochimica Acta* 58(5291–5298).

Ayliffe, L. K., A. M. Lister, and A. R. Chivas
1992 The preservation of glacial-interglacial climate signatures in the oxygen isotopes of elephant skeletal phosphate. *Palaeogeography, Plaeoclimatology, Palaeoecology* 99:179–191.

Badrian, N., and R. Malenky
1984 Feeding ecology of Pan paniscus in the Lonako Forest, Zaire. In *Pygmy chimpanzee*, edited by R. Susman, pp. 275–299. Plenum Press, New York.

Baldwin, P., W. McGrew, and C. Tutin
1982 Wide-ranging chimpanzees at Mt. Asserik, Senegal. *International Journal of Primatology* 3:367–385.

Barry, J.
1995 Faunal turnover and diversity in the terrestrial Neogene of Pakistan. In *Paleoclimate and Evolution with Emphasis on Human Origins*, edited by E. Vrba, G. Denton, T. Partridge and L. Burckle, pp. 275–299. Yale University Press, New Haven.

Barry, J., L. Flynn, and D. Pilbeam
1990 Faunal diversity and turnover in a Miocene terrestrial sequence. In *Causes of Evolution: A Paleontological Perspective*, edited by R. Ross and W. Allmon, pp. 381–421. University of Chicago Press, Chicago.

Barry, J., M. Morgan, L. Flynn, D. Pilbeam, A. Behrensmeyer, S. Raza, I. Khan, C. Badgely, J. Hicks, and J. Kelley
2002 Faunal and environmental change in the Late Miocene Siwaliks of Northern Pakistan. *Paleobiology* (Supplement):1–71.

Barry, J., M. Morgan, L. Flynn, D. Pilbeam, L. Jacobs, E. Lindsay, S. Raza, and N. Solounias
1995 Patterns of faunal turnover and diversity in the Neogene Siwaliks of northern Pakistan. *Palaeogeography, Palaeoclimatology, Palaeoecology* 114:209–226.

Bazzaz, F.
1990 The response of natural ecosystems to the rising of global CO_2 levels. *Annual Review of Ecology and Systematics* 21:167–196.

Behrensmeyer, A., B. Willis, and J. Quade
1995 Floodplains and paleosols of Pakistan Neogene and Wyoming Paleogene deposits: a comparative study. *Palaeogeography, Palaeoclimatology, Palaeoecology* 115:37–60.

Bender, M. M.
1971 Variations in the $^{13}C/^{12}C$ ratios of plants in relation to the pathway of photosynthetic carbon dioxide fixation. *Phytochemistry* 10:1239–1245.

Benton, M.
1990 *Vertebrate Paleontology*. Unwin Hyman, London.

Berggren, W., D. Kent, J. Flynn, and J. Van Couvering
1985 Cenozoic geochronology. *Geological Society of America Bulletin* 96:1407–1418.

Beynon, A., C. Clayton, R. Rozzi, and D. Reid
1998 Radiographic and histological methodologies in estimating the chronology of crown development in modern humans and great apes: a review, with some applications for studies on juvenile hominids. *Journal of Human Evolution* 35:351–370.

Bocherens, H., P. Koch, A. Mariotti, D. Geraads, and J. Jaeger
1996 Isotopic biogeochemistry (^{13}C, ^{18}O) of mammalian enamel from African Pleistocene hominid sites. *Palaios* 11:306–318.

Boyle, E.
1990 Quaternary deepwater paleoceanography. *Science* 249:863–870.

Bryant, J., P. Froelich, W. Showers, and B. Genna
1996a Biological and climatic signals in the oxygen isotopic composition of Eocene-Oligocene equid enamel phosphate. *Palaeogeography, Palaeoclimatology, Palaeoecology* 26:75–89.
1996b A tale of two quarries: biologic and taphonomic signatures in the oxygen isotopic composition of tooth enamel phosphate from modern and Miocene equids. *Palaios* 11:397–408.

Bryant, J., B. Luz, and P. Froelich
1994 Oxygen isotopic composition of fossil horse tooth phosphate as a record of continental paleoclimate. *Palaeogeography, Palaeoclimatology, Palaeoecology* 107:303–316.

Cande, S., and D. Kent
1995 Revised calibration of the geomagnetic polarity timescale for the Late Cretaceous and Cenozoic. *Journal of Geophysical Research* 100B:6093–6095.

Cannon, C., and M. Leighton
1994 Comparative locomotor ecology of gibbons and macaques: selection of canopy elements for crossing gaps. *American Journal of Physical Anthropology* 93:505–524.

1996 Comparative locomotor ecology of gibbons and macaques: does brachiation minimize travel costs? *Tropical Biodiversity* 3:261–267.

Cerling, T. E., and J. M. Harris
1999 Carbon isotope fractionation between diet and bioapatite in ungulate mammals and implications for ecological and paleoecological studies. *Oecologia* 120(3):347–363.

Cerling, T. E., J. M. Harris, S. H. Ambrose, M. G. Leakey, and N. Solounias
1997a Dietary and environmental reconstruction with stable isotope analyses of herbivore tooth enamel from the Miocene locality of Fort Ternan, Kenya. *Journal of Human Evolution* 33:635–650.

Cerling, T. E., J. M. Harris, B. J. MacFadden, M. G. Leakey, J. Quade, V. Eisenmann, and J. R. Ehleringer
1997b Global vegetation change through the Miocene/Pliocene boundary. *Nature* 389(6647):153–158.

Cerling, T. E., J. M. Harris, and B. H. Passey
2003 Dietary preferences of East African Bovidae based on stable isotope analysis. *Journal of Mammalogy* (in press).

Cerling, T., J. Quade, Y. Wang, and J. Bowman
1989 Carbon isotopes in soils and paleosols as ecological and paleoecological indicators. *Nature* 341:138–139.

Cerling, T., and Z. Sharp
1996 Stable carbon and oxygen isotope analysis of fossil horse tooth enamel using laser ablation. *Palaeogeography, Palaeoclimatology, Palaeoecology* 126:173–186.

Cerling, T., Y. Wang, and J. Quade
1993 Expansion of C4 ecosystems as an indicator of global ecological change in the late Miocene. *Nature* 361:344–345.

Chivers, D.
1984 Feeding and ranging in gibbons: a summary. In *The Lesser Apes*, edited by H. Preuschoft, D. Chivers, W. Brockelman and N. Creel, pp. 267–281. Edinburgh University Press, Edinburgh.

Corvinus, G., and L. Rimal
2001 Biostratigraphy and geology of the Neogene Siwalik Group of the Surai Khola and Rato Khola areas in Nepal. *Palaeogeography, Palaeoclimatology, Palaeoecology* 165:251–279.

Curran, L., I. Caniago, G. Paoli, D. Astianti, M. Kusneti, M. Leighton, C. Nirarita, and H. Haeruman
1999 Impact of El Nino and logging on canopy tree recruitment in Borneo. *Science* 286:2184–2188.

Dansgaard, W.
1964 Stable isotopes in precipitation. *Tellus* 16:436–468.

DeLucia, E., J. Hamilton, S. Naidu, R. Thomas, J. Andrews, A. Finzi, M. Lavine, R. Matamala, J. Mohan, G. Hendrey, and W. Schlesinger
1999 Net primary production of a forest ecosystem with experimental CO_2 enrichment. *Science* 284:1177–1179.

DeNiro, M., and E. S. Epstein
1978 Influence of diet on the distribution of carbon isotopes in animals. *Geochimica et Cosmochimica Acta* 42:495–506.

Dettman, D. L., M. J. Kohn, J. Quade, F. J. Ryerson, T. P. Ojha, and S. Hamidullah
2001 Seasonal stable isotope evidence for a strong Asian monsoon throughout the past 10.7 m.y. *Geology* 29:31–34.

Diamond, J.
1984 "Normal" extinctions of isolated populations. In *Extinctions*, edited by M. Nitecki, pp. 191–246. University of Chicago Press, Chicago.

Dongmann, G., H. Nurnberg, H. Forstel, and K. Wagener
1974 On the enrichment of H218O in the leaves of transpiring plants. *Radiation and Environmental Biophysics* 11:41–52.

Ehleringer, J., T. Cerling, and B. Helliker
1997 C4 photosynthesis, atmospheric CO2, and climate. *Oecologia* 112:285–299.

Ehleringer, J. R., C. B. Field, Z. F. Lin, and C. Y. Kuo
1986 Leaf carbon isotope and mineral-composition in subtropical plants along an irradiance cline. *Oecologia* 70(4):520–526.

Ehleringer, J. R., Z. F. Lin, C. B. Field, G. C. Sun, and C. Y. Kuo
1987 Leaf carbon isotope ratios of plants from a subtropical monsoon forest. *Oecologia* 72(1):109–114.

Ehleringer, J. R., R. F. Sage, L. B. Flanagan, and R. W. Pearcy
1991 Climate change and the evolution of C4 photosynthesis. *Trends in Ecology and Evolution* 6:95–99.

Emmons, L. H., A. Gautier-Hion, and G. Dubost
1983 Community structure of the frugivorous-folivorous mammals of Gabon. *Journal of Zoology*, London 199:209–222.

Epstein, S., P. Thompson, and C. Yapp
1977 Oxygen and hydrogen isotopic ratios in plant cellulose. *Science* 198:1209–1215.

Farquhar, G.
1997 Carbon dioxide and vegetation. *Science* 278:1411.

Farquhar, G. D., J. R. Ehleringer, and K. T. Hubick
1989 Carbon isotope discrimination and photosynthesis. *Annual Review of Plant Physiology and Plant Molecular Biology* 40:503–537.

Filippelli, G.
1997 Intensification of the Asain monsoon and a chemical weathering event in the late Miocene-early Pliocene: implications for late Neogene climate change. *Geology* 25:27–30.

Fitzgerald, C.
1998 Do enamel microstructures have regular time dependency? Conclusions from the literature and a large-scale study. *Journal of Human Evolution* 35:371–386.

Fleming, T., R. Breitwisch, and G. Whitesides
1987 Patterns of tropical vertebrate fruigvore diversity. *Annual Review of Ecology and Systematics* 18:91–109.

Flynn, L.
1982 Systematic revision of the Siwalik *Rhyzomidae* (Rodentia). *Geobios* 15:327–389.

Flynn, L., J. Barry, M. Morgan, D. Pilbeam, L. Jacobs, and E. Lindsay
1995 Neogene Siwalik mammalian lineages: species longevities, rates of change, and modes of speciation. *Palaeogeography, Palaeoclimatology, Palaeoecology* 114:249–264.

Flynn, L., and L. Jacobs
1982 Effects of changing environments on Siwalik rodent faunas of northern Pakistan. *Palaeogeography, Palaeoclimatology, Palaeoecology* 38:129–138.

Fontes, J., R. Gonfiantini, and M. Roche
1970 Deuterium et oxygene-18 dans les eaux du lac Tchad. In *Isotope Hydrology*, pp. 387. International Atomic Energy Agency, Vienna.

Fox, D., and D. Fisher
2001 Stable isotope ecology of a Late Miocene population of Gomphotherium productus (Mammalia, Proboscidea) from Port of Entry Pit, Oklahoma, USA. *Palaios* 16:279–293.

Fricke, H., W. Clyde, J. O'Neil, and P. Gingerich
1998 Evidence for rapid climate change in North America during the latest Paleocene thermal maximum: oxygen isotope compositions of biogenic phosphate form the Bighorn Basin (Wyoming). *Earth and Planetary Science Letters* 160:193–208.

Fricke, H., and J. O'Neil
1996 Inter- and intra-tooth variation in the oxygen isotope composition of mammalian tooth enamel phosphate: implications for paleoclimatological and paleobiological research. *Palaeogeography, Palaeoclimatology, Palaeoecology* 126:91–99.

Fritz, P.
1981 River waters. In *Stable Isotope Hydrology Deuterium and Oxygen-18 in the Water Cycle*, edited by J. Gat and R. Gonfiantini, pp. 177–201. Technical Report Series. vol. 210. International Atomic Energy Agency, Vienna.

Gadbury, C., L. Todd, A. Jahren, and R. Amundson
2000 Spatial and temporal variations in the isotopic composition of bison tooth enamel from the Early Holocene Hudson-Meng Bone Bed, Nebraska. *Palaeogeography, Palaeoclimatology, Palaeoecology* 157:79–93.

Gat, J.
1981 Groundwater. In *Stable Isotope Hydrology Deuterium and Oxygen-18 in the Water Cycle*, edited by J. Gat and R. Gonfiantini, pp. 223–239. Technical Report Series. vol. 210. International Atomic Energy Agency, Vienna.

Gautier-Hion, A., L. H. Emmons, and G. Dubost
1980 A comparison of the diets of three major groups of primary consumers of Gabon (primates, squirrels, and ruminants). *Oecologia* 45:182–189.

Gittens, S., and R. Tilson
1984 Notes on the ecology and behaviour of the hoolock gibbon. In *The Lesser Apes*, edited by H. Preuschoft, D. Chivers, W. Brockelman and N. Creel, pp. 258–266. Edinburgh University Press, Edinburgh.

Gonfiantini, R.
1985 On the isotopic composition of precipitation in tropical stations. *Acta Amazonica* 15:121–139.

Goodall, J.
1968 The behaviour of free-ranging chimpanzees in the Gombe Stream Reserve. *Animal Behaviour Monographs* 1:161–311.
1986 The *Chimpanzees of Gombe*. Belknap Press, Cambridge.

Gordon, K.
1982 A study of microwear on chimpanzee molars: implications for dental microwear analysis. *American Journal of Physical Anthropology* 59:195–215.

Grine, F., and R. Kay
1988 Early hominid diets from quantitative image analysis of dental microwear. *Nature* 333:765–768.

Guo, Z., W. Ruddiman, Q. Hao, H. Wu, Y. Qiao, R. Zhu, S. Peng, B. Wei, B. Yuan, and T. Liu
2002 Onset of Asian desertification by 22 Myr ago inferred from loess deposits in China. *Nature* 416:159–163.

Harris, J., and T. Cerling
2002 Dietary adaptations of extant and Neogene African suids. *Journal of Zoology*, London 256:45–54.

Harrison, J.
1962 The distribution of feeding habits among animals in a tropical rain forest. *Journal of Animal Ecology* 31:53–63.

Heydon, M., and P. Bulloh
1997 Mousedeer densities in a tropical rainforest: the impact of selective logging. *Journal of Applied Ecology* 34:484–496.

Hillson, S.
1986 *Teeth*. Cambridge University Press, Cambridge.

Hladik, C.
1977 *Chimpanzees of Gabon and chimpanzees of Gombe: some comparative data on the diet*. In *Primate Ecology*, edited by T. Clutton-Brock, pp. 481–501. Academic Press, New York.

IAEA
1981 *World Survey of Isotope Concentrations in Precipitation*. Technical Report Series 206. International Atomic Energy Agency, Vienna.

IAEA/WMO
2001 Global Network for Isotopes in Precipitation. *The GNIP Database*. Accessible at: http://isohis.iaea.org.

Jablonski, N., M. Whitfort, N. Roberts-Smith, and X. Qinqi
2000 The influence of life history and diet on the distribution of catarrhine primates during the Pleistocene in eastern Asia. *Journal of Human Evolution* 39:131–157.

Janis, C.
1982 Evolution of horns in ungulates: ecology and paleoecology. *Biological Reviews* 57:261–318.
1990 Correlation of reproductive and digestive strategies in the evolution of cranial appendages. In *Horns, Pronghorns, and Antlers Evolution, Morphology, Physiology, and Social Significance*, edited by G. Bubenik, and A. Bubenik, pp. 114–133. Springer-Verlag, New York.

Janis, C., J. Damuth, and J. Theodor
2000 Miocene ungulates and terrestrial primary productivity: where have all the browsers gone? *Proceedings of the National Academy of Sciences* 97:7899–7904.
2002 The origins and evolution of the North American grassland biome: the story from the hoofed mammals. *Palaeogeography, Palaeoclimatology, Palaeoecology* 177:183–198.

Johns, A., and J. Skorupa
1987 Responses of rain-forest primates to habitat disturbance: a review. *International Journal of Primatology* 8:157–191.

Kano, T.
1972 Distribution and adaptation of the chimpanzee on the eastern shore of Lake Tanganyika. *Kyoto University African Studies* 7:37–129.

Kappeler, M.
1984 The gibbon in Java. In *The Lesser Apes*, edited by H. Preushchoft, D. Chivers, W. Brockelman, and N. Creel, pp. 19–31. Edinburgh University Press, Edinburgh.

Kay, R.
1977 The evolution of molar occlusion in Cercopithecoidea and early catarrhines. *American Journal of Physical Anthropology* 46:327–352.
1981 The nut-crackers—a new theory of the adaptations of the Ramapithecinae. *American Journal of Physical Anthropology* 55:141–151.

Kay, R., and K. Hiiemae
1974 Jaw movement and tooth use in recent and fossil primates. *Americal Journal of Physical Anthropology* 40:227–256.

Kelley, J.
1986 Paleobiology of Miocene Hominoids. Ph.D. dissertation, Yale University.

1997 Paleobiological and phylogenetic significance of like history in Miocene hominoids. In *Function, Phylogeny and Fossils: Miocene Hominoid Origins*, edited by D. Begun, C. Ward and M. Rose, pp. 173–208. Plenum Press, New York

Kelley, J., and D. Pilbeam
1986 The Dryopithecines: taxonomy, comparative anatomy, and phylogeny of Miocene large hominoids. In *Comparative Primate Biology vol. 1: Systematics, Evolution, and Anatomy*, edited by D. Swindler, and J. Erwin, pp. 361–411. Alan R. Liss, Inc., New York.

Kennett, J.
1995 A review of polar climatic evolution during the Neogene, based on the marine sediment record. In *Paleoclimate and Evolution with Emphasis on Human Origins*, edited by E. Vrba, G. Denton, T. Partridge, and L. Burckle, pp. 49–64. Yale Univesity Press, New Haven.

King, T.
2001 Dental microwear and diet in Eurasian Miocene catarrhines. In *Phylogeny of the Neogene Hominoid Primates of Eurasia*, vol. 2, edited by L. de Bonis, G. D. Koufos, and P. Andrews, pp. 102–117. Cambridge University Press, Cambridge.

King, T., L. Aiello, and P. Andrews
1999 Dental microwear of *Griphopithecus alpani*. *Journal of Human Evolution* 36:3–31.

Kingdon, J.
1997 *The Kingdon Field Guide to African Mammals*. Academic Press, San Diego.

Knott, C.
1998 Changes in organgutan caloric intake, energy balance, and ketones in response to fluctuating fruit availability. *International Journal of Primatology* 19:1061–1079.

Koch, C.
1987 Prediction of sample size effects on the measured temporal and geographic distribution patterns of species. *Paleobiology* 13:100–107.

Koch, P., D. Fisher, and D. Dettman
1989 Oxygen isotope variation in the tusks of extinct proboscideans: a measure of season of death and seasonality. *Geology* 17:515–519.

Koch, P. L., J. C. Zachos, and D. L. Dettman
1997 The effects of sample treatment and diagenesis on the isotopic integrity of carbonate in biogenic hydroxylapatite. *Journal of Archaeological Science* 24:417–429.

Kohn, M., M. Schoeninger, and J. Valley
1996 Herbivore tooth oxygen isotope compositions: effects of diet and physiology. *Geochimica et Cosmochimica Acta* 60:3889–3896.

Kortlandt, A.
1983 Marginal habitats of chimpanzees. *Journal of Human Evolution* 12:231–278.

Kroon, D., T. Steens, and S. R. Troelstra
1991 Onset of monsoonal related upwelling in the western Arabian Sea as revealed by planktonic foraminifers. *Proceedings of the Ocean Drilling Program, Scientific Results* 117:257–263.

Lambert, F.
1989 Fig-eating by birds in a Malaysian lowland rain-forest. *Journal of Tropical Ecology* 5:401–412.

Lee-Thorp, J. A., N. J. Van der Merwe, and C. K. Brain
1989 Isotopic evidence for dietary differences between two extinct baboon species from Swartkrans. *Journal of Human Evolution* 18:183–190.

Leighton, M.
1993 Modeling dietary selectivity by Bornean orangutans: evidence for integration of multiple criteria in fruit selection. *International Journal of Primatology* 14:257–313.

Longinelli, A.
1984 Oxygen isotopes in mammal bone phosphate: A new tool for paleohydrological and paleoclimatological research? *Geochimica et Cosmochimica Acta* 48:385–390.

Luz, B., and Y. Kolodny
1985 Oxygen isotope variations in phosphates of biogenic apatites, IV: Mammal teeth and bones. *Earth and Planetary Science Letters* 75:29–36.

Luz, B., Y. Kolodny, and M. Horowitz
1984 Fractionation of oxygen isotopes between mammalian bone-phosphate and environmental drinking water. *Geochimica et Cosmochimica Acta* 48:1689–1693.

MacKinnon, J.
1974 The behaviour and ecology of wild orangutans (*Pongo pygmaeus*). *Animal Behaviour* 22:3–74.
1977 A comparative ecology of Asian apes. *Primates* 18:747–772.

MacKinnon, J., and K. MacKinnon
1978 Comparative feeding ecology of six sympatric primates in West Malaysia. In *Recent Advances in Primatology*, edited by D. Chivers and J. Herbert, pp. 305–321. vol. 1. Academic Press, London.

Malenky, R., and R. Wrangham
1994 A quantitative comparison of terrestrial herbaceous food consumption by *Pan paniscus* in the Lomako Forest, Zaire, and *Pan troglodytes* in the Kibale Forest, Uganda. *American Journal of Primatology* 32:1–12.

McGee, V.
1971 *Principles of Statistics*. Appleton-Century-Crofts, New York.

McGrew, W., P. Baldwin, and C. Tutin
1981 Chimpanzees in a hot, dry open habitat: Mt. Asserik, Senegal, West Africa. *Journal of Human Evolution* 10:227–244.

Medina, E., G. Montes, E. Cuevas, and Z. Rokzandic
1986 Profiles of CO_2 concentration and $\delta^{13}C$ values in tropical rain forests of the upper Rio Negro Basin, Venezuela. *Journal of Tropical Ecology* 2:207–217.

Mooley, D., and J. Shukla
1987 Variability and forecasting of the summer monsoon rainfall over India. In *1987*, edited by C.-P. Chang and T. Krishnamurti, pp. 26–59. Oxford Monographs on Geology and Geophysics, vol. 7. Oxford University Press, Oxford.

Morgan, B.
1968 *An Introduction to Bayesian Statistical Decision Process*. Prentice-Hall, Englewood Cliffs.

Morgan, M.
1994 Paleoecology of Siwalik Miocene Hominoid Communities: Stable Carbon Isotope, Dental Microwear, and Body Size Analyses. Ph.D., Harvard University.

Morgan, M., J. Kingston, and B. Marino
1994 Carbon isotopic evidence for the emergence of C4 plants in the Neogene from Pakistan and Kenya. *Nature* 367:162–165.

Murakami, T.
1987 *Effects of the Tibetan Plateau*. In *Monsoon Meteorology*, edited by C.-P. Chang and T. Krishnamurti, pp. 235–270. Oxford Monographs on Geology and Geophysics, vol. 7. Oxford University Press, Oxford.

Newmark, W.
1985 Legal and biotic boundaries of Western North American National Parks: a problem of congruence. *Biological Conservation* 33:197–208.
1987 A land-bridge island perspective on mammalian extinctions in western North American parks. *Nature* 325:430–432.
1996 Insularization of Tanzanian parks and the local extinction of large mammals. *Conservation Biology* 10:2549–1556.

Nishida, T., and K. Kawanaka
1972 Inter-unit-group relationships among wild chimpanzees of the Mahali Mountains. *Kyoto University African Studies* 7:131–169.

Njitchoua, R., L. Sigha-Nkamdjou, L. Dever, C. Marlin, D. Sighomnou, and P. Nia
1999 Variations of the stable isotopic compositions of rainfall events from the Cameroon rain forest, Central Africa. *Journal of Hydrology* 223:17–26.

Nowack, R.
1999 *Walker's Mammals of the World* 2. Johns Hopkins University Press, Baltimore.

Olkin, I., and R. Tate
1961 Multivariate correlation models with mixed discrete and continuous-variables. *Annals of Mathematical Statistics* 32:448–465.

Pagani, M., K. Freeman, and M. Arthur
1999 Late Miocene atmospheric CO_2 concentrations and the expansion of C_4 grasses. *Science* 285:876–879.

Pearson, P. and M. Palmer
2000 Atmospheric carbon dioxide concentrations over the past 60 million years. *Nature* 406:695–699.

Prasad, M.
1993 Siwalik (Middle Miocene) woods from the Kalagarh area in the Himalayan foot hills and their bearing on paleoclimate and phytogeography. *Review of Paleobotany and Palynology* 76:49–82.

Prell, W., and J. Kutzbach
1992 Sensitivity of the Indian monsoon to forcing parameters and implications for its evolution. *Nature* 360:647–652.

Quade, J., and T. Cerling
1995 Expansion of C4 grasslands in the Late Miocene of Northern Pakistan—evidence from stable isotopes in paleosols. *Palaeogeography, Palaeoclimatology, Palaeoecology* 115:91–116.

Quade, J., T. Cerling, and J. Bowman
1989 Development of Asian monsoon revealed by marked ecological shift during the latest Miocene in northern Pakistan. *Nature* 342:163–165.

Quade, J., T. E. Cerling, J. C. Barry, M. E. Morgan, D. R. Pilbeam, A. R. Chivas, J. A. Leethorp, and N. J. Vandermerwe
1992 A 16-Ma record of paleodiet using carbon and oxygen isotopes in fossil teeth from Pakistan. *Chemical Geology* 94(3):183–192.

Quade, J., J. Cater, T. Ojha, J. Adam, and T. Harrison
1995a Late Miocene environmental change in Nepal and the northern Indian subcontinent: stable isotopic evidence from paleosols. *GSA Bulletin* 107:1381–1397.

Quade, J., T. E. Cerling, P. Andrews, and B. Alpagut
1995b Paleodietary reconstruction of Miocene faunas from Pasalar, Turkey using stable carbon and oxygen isotopes of fossil tooth enamel. *Journal of Human Evolution* 28:373–384.

Raymo, M., and W. Ruddiman
1992 Tectonic forcing of late Cenozoic climate. *Nature* 359:117–122.

Rea, D.
1992 Delivery of Himalayan sediment to the northern Indian Ocean and its relation to global climate, sea level, uplift, and seawater strontium. In *Synthesis of Results from Scientific Drilling in the Indian Ocean*, edited by R. Duncan, pp. 387–402. America Geophysical Union Geophysical Monograph, vol. 70.

Retallack, G.
2001 A 300-million-year record of atmospheric carbon dioxide from fossil plant cuticles. *Nature* 411:287–290.

Richards, P.
1996 *The Tropical Rainforest*. Cambridge University Press, Cambridge.

Rodman, P.
1977 Feeding behavior of orangutans in the Kutai Reserve, East Kalimantan. In *Primate Ecology*, edited by T. Clutton-Brock, pp. 383–413. Academic Press, London.

Rose, M.
1986 Further hominoid postcranial specimens from the Late Miocene Nagri formation of Pakistan. *Journal of Human Evolution* 15:333–368.

Royer, D., S. Wing, D. Beerling, D. Jolley, P. Koch, L. Hickey, and R. Berner
2001 Paleobotanical evidence for near present-day levels of atmospheric CO2 during part of the Tertiary. *Science* 292:2310–2313.

Rozanski, K., L. Arguas-Araguas, and R. Gonfiantini
1993 Isotopic patterns in modern global precipitation. In *Climate Change in Continental Isotope Records*, edited by P. Swart, pp. 1–36. American Geophysical Union, Washington.

Sharp, Z., and T. Cerling
1996 A laser GC-IRMS technique for *in situ* stable isotope analyses of carbonates and phosphates. *Geochimica et Cosmochimica Acta* 60:2909–2916.
1998 Fossil isotope records of seasonal climate and ecology: straight from the horse's mouth. *Geology* 26:219–222.

Shellis, R.
1998 Utilization of periodic markings in enamel to obtain information on tooth growth. *Journal of Human Evolution* 35:387–400.

Sigman, D., and E. Boyle
2000 Glacial/interglacial variations in atmospheric carbon dioxide. *Nature* 407:859–869.

Skorupa, J.
1986 Responses of rainforest primates to selective loggin in Kibale Forest, Uganda: a summary report. In *Primates: the Road to Self-Sustaining Populations*, edited by K. Benirschke, pp. 57–70. Springer Verlag, New York.

Smith, B.
1989 Dental development as a measure of life history in primates. *Evolution* 43:683–688.
1991 Dental development and the evolution of life history in Hominidae. *American Journal of Physical Anthropology* 86:157–174.

Smith, S., T. Huxman, S. Zitzer, T. Charlet, D. Housman, J. Coleman, L. Fenstermaker, J. Seemann, and R. Nowak
2000 Elevated CO2 increases productivity and invasive species success in an arid ecosystem. *Nature* 408:79–82.

Solounias, N., and L. Hayek
1993 New methods of tooth microwear analysis and application to dietary determination of two extinct antelopes. *Journal of Zoology*, London 229:421–445.

Solounias, N., and G. Semprebon
2002 Advances in the reconstruction of ungulate ecomorphology and application to early fossil equids. *American Museum Novitates* 3366.

Solounias, N., M. Teaford, and A. Walker
1988 Interpreting the diet of extinct ruminants:the case of a non-browsing giraffid. *Paleobiology* 14:287–300.

Stern, L., G. Johnson, and C. Chamberlain
1994 Carbon isotope signature of environmental change found in fossil ratite eggshells from a South Asian Neogene sequence. *Geology* 22:419–422.

Sternberg, L., M. DeNiro, and H. Johnson
1984 Isotope ratios of cellulose from plants having different photosynthetic pathways. *Plant Physiology* 74:557–561.

Tanner, M., and W. Wong
1987 An application of imputation to an estimation problem in grouped lifetime analysis. *Technometrics* 29:23–32.

Teaford, M., and O. Oyen
1989 In vivo and in vitro turnover in dental microwear. *American Journal of Physical Anthropology* 80:447–460.

Teaford, M., and A. Walker
1984 Quantitative differences in dental microwear between primate species with different diets and a comment on the presumed diet of Sivapithecus. *American Journal of Physical Anthropology* 64:191–200.

Terborgh, J.
1990 An Overview of research at Cocha Cashu Biological Station. In *Four Neotropical Forests*, edited by A. Gentry, pp. 48–59. Yale University Press, New Haven.

Tutin, C., M. Fernandez, M. Rogers, E. Williamson, and W. McGrew
1991 Foraging profiles of sympatric lowland gorillas and chimpanzees in the Lope Reserve, Gabon. *Philosophical Transactions of the Royal Society of London* B 334:179–186.

Ungar, P.
1996 Dental microwear of European Miocene catarrhines: evidence for diets and tooth use. *Journal of Human Evolution* 30:335–366.

Van der Merwe, N. J.
1982 Carbon isotopes, photosynthesis, and archaeology. *American Scientist* 70:596–606.

Van der Merwe, N. J., and E. Medina
1989 Photosynthesis and $^{13}C/^{12}C$ ratios in Amazonian rain forests. *Geochimica et Cosmochimica Acta* 53:1091–1094.
1991 The canopy effect, carbon isotope ratios and foodwebs in Amazonia. *Journal of Archaeological Science* 18:249–259.

Van der Merwe, N. J., and J. C. Vogel
1978 ^{13}C content of human collagen as a measure of prehistoric diet in woodland North America. *Nature* 276:815–816.

Vogel, J. C.
1980 *Fractionation of the Carbon Isotopes During Photosynthesis*. Springer-Verlag, Berlin.

Walker, A., H. Hoeck, and L. Perez
1978 Microwear of mammalian teeth as an indicator of diet. *Science* 201:908–910.

Walter, H.
1973 *Vegetation of the Earth in Relation to Climate and the Eco-physiological Conditions*. Springer-Verlag, New York.

Walter, H., and S. Breckle
1984 *Ecological Systems of the Geobiosphere 2 Tropical and Subtropical Zonobiomes 2.* Springer-Verlag, New York.

Walter, H., E. Harnickell, and D. Mueller-Dombois
1975 *Climate Diagram Maps.* Springer-Verlag, New York.

Wang, Y., and T. E. Cerling
1994 A model of fossil tooth and bone diagenesis—implications for paleodiet reconstruction from stable isotopes. *Palaeogeography, Palaeoclimatology, Palaeoecology* 107(3–4):281–289.

Webster, P.
1987 The elementary monsoon. In *Monsoons,* edited by J. Fein and P. Stephens, pp. 3–32. John Wiley and Sons, New York.

Wich, S., and C. Van Schaik
2000 The impact of El Nino on mast fruiting in Sumatra and elsewhere in Malesia. *Journal of Tropical Ecology* 16:563–577.

Willis, B., and A. Behrensmeyer
1995 Fluvial systems in the Siwalik Miocene and Wyoming Paleogene. *Palaeogeography, Palaeoclimatology, Palaeoecology* 115:13–35.

Wrangham, R.
1975 The Behavioural Ecology of Chimpanzees in Gombe National Park, Tanzania. Ph.D., University of Camridge.

Wrangham, R., N. Conklin, C. Chapman, and K. Hunt
1991 The significance of fibrous foods for Kibale Forest chimpanzees. *Philosophical Transactions of the Royal Society of London* B 334:171–178.
1998 Dietary response of chimpanzees and cercopithecines to seasonal variation in fruit abundance I. Antifeedants. *International Journal of Primatology* 19:949–969.

Wrangham, R., M. Rogers, and G. Basuta
1993 Ape food density in the ground layer in Kibale Forest, Uganda. *African Journal of Ecology* 31:49–57.

Yurtsever, Y., and J. Gat
1981 Atmospheric Waters. *Stable Isotope Hydrology: IAEA Technical Report Series* 210:103–142.

Zazzo, A., H. Bocherens, M. Brunet, A. Beauvilain, B. D., H. Mackaye, P. Vignaud, and A. Mariotti
2000 Herbivore paleodiet and paleoenvironmental changes in Chad during the Pliocene using stable isotope ratios of tooth enamel carbonate.

Index